Healthy Worker and Healthy Organization

Occupational Safety, Health, and Ergonomics: Theory and Practice

Series Editor: Danuta Koradecka

(Central Institute for Labour Protection – National Research Institute)

This series will contain monographs, references, and professional books on a compendium of knowledge in the interdisciplinary area of environmental engineering, which covers ergonomics and safety and the protection of human health in the working environment. Its aim consists in an interdisciplinary, comprehensive and modern approach to hazards, not only those already present in the working environment, but also those related to the expected changes in new technologies and work organizations. The series aims to acquaint both researchers and practitioners with the latest research in occupational safety and ergonomics. The public, who want to improve their own or their family's safety, and the protection of heath will find it helpful, too. Thus, individual books in this series present both a scientific approach to problems and suggest practical solutions; they are offered in response to the actual needs of companies, enterprises, and institutions.

Individual and Occupational Determinants: Work Ability in People with Health Problems
Joanna Bugajska, Teresa Makowiec-Dąbrowska, Tomasz Kostka

Healthy Worker and Healthy Organization: A Resource-Based Approach
Dorota Żołnierczyk-Zreda

Emotional Labour in Work with Patients and Clients: Effects and Recommendations for Recovery
Dorota Żołnierczyk-Zreda

New Opportunities and Challenges in Occupational Safety and Health Management
Daniel Podgórski

Emerging Chemical Risks in the Work Environment
Małgorzata Pośniak

Visual and Non-Visual Effects of Light: Working Environment and Well-Being
Agnieszka Wolska, Dariusz Sawicki, Małgorzata Tafil-Klawe

Occupational Noise and Workplace Acoustics: Advances in Measurement and Assessment Techniques
Dariusz Pleban

Virtual Reality and Virtual Environments: A Tool for Improving Occupational Safety and Health
Andrzej Grabowski

Head, Eye, and Face Personal Protective Equipment: New Trends, Practice and Applications
Katarzyna Majchrzycka

Nanoaerosols, Air Filtering and Respiratory Protection: Science and Practice
Katarzyna Majchrzycka

Microbial Corrosion of Buildings: A Guide to Detection, Health Hazards, and Mitigation
Rafał L. Górny

Respiratory Protection Against Hazardous Biological Agents
Katarzyna Majchrzycka, Małgorzata Okrasa, Justyna Szulc

For more information about this series, please visit: https://www.crcpress.com/Occupational-Safety-Health-and-Ergonomics-Theory-and-Practice/book-series/CRCOSHETP

Healthy Worker and Healthy Organization

A Resource-Based Approach

Edited by
Dorota Żołnierczyk-Zreda

CRC Press
Taylor & Francis Group
Boca Raton London New York

CRC Press is an imprint of the
Taylor & Francis Group, an **informa** business

First edition published 2021
by CRC Press
6000 Broken Sound Parkway NW, Suite 300, Boca Raton, FL 33487-2742

and by CRC Press
2 Park Square, Milton Park, Abingdon, Oxon, OX14 4RN

CRC Press is an imprint of Taylor & Francis Group, an Informa business

Visit the Taylor & Francis Web site at
http://www.taylorandfrancis.com

and the CRC Press Web site at
http://www.crcpress.com

ISBN: 978-0-367-86060-8 (hbk)
ISBN: 978-1-003-03243-4 (ebk)

Typeset in Times
by Deanta Global Publishing Services, Chennai, India

Contents

Contents

Acknowledgment

Wishing a special thank you to Paulina Barańska for translating this book into English.

Editor

Dorota Żołnierczyk-Zreda, PhD, Head of Laboratory of Social Psychology, Central Institute for Labour Protection – National Research Institute, Warsaw, Poland, is a senior researcher in occupational health psychology. Her research is focused on psychosocial working conditions and mental health at work, its determinants and various methods to sustain it in different groups of workers (e.g. young, older) and different occupational groups. She has extensive experience in qualitative studies and quantitative studies as well as in policy research and monitoring studies. She is also involved in national and international projects investigating different stress management interventions on both the organizational and individual levels. She is the author or co-author of approximately 60 scientific publications, including articles, chapters in monographs and textbooks, and many speeches at scientific conferences nationally and abroad. She is a licensed cognitive-behavioral therapist.

Contributors

Łukasz Baka, PhD, psychologist, is a senior researcher working in the Department of Ergonomics, Central Institute for Labour Protection – National Research Institute, Warsaw, Poland. His research interests focus on the problems of job stress and its health- and organization-related implications: burnout, work engagement, and counterproductive work behavior. He has published three books and several articles on that subject.

Katarzyna Hildt-Ciupińska, PhD, is a researcher working in the Department of Ergonomics, Central Institute for Labour Protection – National Research Institute, Warsaw, Poland. Her main activities include research and development in the area of work-life balance, health and safety at work among employees, health promotion, healthy lifestyle and health behaviors of employees, as well as among older people and people with disabilities.

Zofia Mockałło, MA psychologist, is a researcher working in the Department of Ergonomics, Laboratory of Social Psychology, Central Institute for Labour Protection – National Research Institute, Warsaw, Poland, and member of a PEROH Wellbeing at Work Initiative. Her research area is focused on work and organizational psychology, sources and effects of employees' wellbeing with special interest in mental health, job and individual resources, and innovation.

Magdalena Warszewska-Makuch, PhD, psychologist, is a researcher working in the Department of Ergonomics, Central Institute for Labour Protection – National Research Institute, Warsaw, Poland, and the Laboratory of Social Psychology. She received her PhD in social psychology in 2015 from the SWPS University of Social Sciences and Humanities in Warsaw, Poland. Her research area is focused on workplace bullying and cyberbullying, on the relationship between psychosocial working conditions, employee participation, ageing, stress at work, and health.

Series Editor

Professor Danuta Koradecka, PhD, D.Med.Sc. and Director of the Central Institute for Labour Protection – National Research Institute (CIOP-PIB), is a specialist in occupational health. Her research interests include the human health effects of hand-transmitted vibration; ergonomics research on the human body's response to the combined effects of vibration, noise, low temperature and static load; assessment of static and dynamic physical load; development of hygienic standards as well as development and implementation of ergonomic solutions to improve working conditions in accordance with International Labour Organisation (ILO) convention and European Union (EU) directives. She is the author of over 200 scientific publications and several books on occupational safety and health.

***

The "Occupational Safety, Health, and Ergonomics: Theory and Practice" series of monographs is focused on the challenges of the 21st century in this area of knowledge. These challenges address diverse risks in the working environment of chemical (including carcinogens, mutagens, endocrine agents), biological (bacteria, viruses), physical (noise, electromagnetic radiation) and psychophysical (stress) nature. Humans have been in contact with all these risks for thousands of years. Initially, their intensity was lower, but over time it has gradually increased, and now too often exceeds the limits of man's ability to adapt. Moreover, risks to human safety and health, so far assigned to the working environment, are now also increasingly emerging in the living environment. With the globalisation of production and merging of labour markets, the practical use of the knowledge on occupational safety, health, and ergonomics should be comparable between countries. The presented series will contribute to this process.

The Central Institute for Labour Protection – National Research Institute, conducting research in the discipline of environmental engineering, in the area of working environment and implementing its results, has summarised the achievements – including its own – in this field from 2011 to 2019. Such work would not be possible without cooperation with scientists from other Polish and foreign institutions as authors or reviewers of this series. I would like to express my gratitude to all of them for their work.

It would not be feasible to publish this series without the professionalism of the specialists from the Publishing Division, the Centre for Scientific Information and Documentation, and the International Cooperation Division of our Institute. The challenge was also the editorial compilation of the series and ensuring the efficiency of this publishing process, for which I would like to thank the entire editorial team of CRC Press – Taylor & Francis.

***

This monograph, published in 2020, has been based on the results of a research task carried out within the scope of the second to fourth stage of the Polish National Programme "Improvement of safety and working conditions" partly supported – within the scope of research and development – by the Ministry of Science and Higher Education/National Centre for Research and Development, and within the scope of state services – by the Ministry of Family, Labour and Social Policy. The Central Institute for Labour Protection – National Research Institute is the Programme's main coordinator and contractor.

1 Introduction

Dorota Żołnierczyk-Zreda

The World Health Organization (WHO) defines health as 'a state of complete physical, mental and social wellbeing, and not merely the absence of disease or infirmity' (WHO 2006). The definition is widely used in occupational health research, as is the *Healthy Workplaces* model also developed by WHO (Hassard et al. 2017). The model encompasses physical, psychological and social aspects of workplace wellbeing, and an additional aspect – *personal health resources*. These include organizational measures supporting the work environment, such as a provision of health services, resources, information and opportunities aimed at motivating employees to maintain or strengthen individual health behaviors. The physical aspect, i.e. the *physical work environment*, introduces safe and healthy physical working conditions covering such factors as equipment, temperature, lighting, noise, body posture and exposure to chemicals. The psychological aspect comprises supportive organizational solutions in time management, social support, the employee's control over his/her work and supervision at the workplace. Equally, the *Healthy Workplaces* model accounts for the social context of organizational operations defined as *enterprise community involvement*. Here, the focus is on the relationship between the enterprise and the community, and the ways to mitigate a potentially negative impact.

Other definitions of a healthy workplace focus on employee well-being, which that should be fostered at the enterprise level, as this results in greater employee involvement and greater productivity, which are key to the company's performance (Cooper et al. 2004; Schmidt et al. 2000; Grawitch et al. 2006). The WHO *Healthy Workplaces* model components are referred to in the literature as 'paths to a healthy workplace' (Grawitch et al. 2006), 'best practices' (Fitz-enz 1993; Van De Voorde et al. 2012) or 'human resource management practices' (Jiang et al. 2012; Guest 2017).

This publication presents research into employee and organizational health and well-being, which can be described in terms of resources, defined by Halbesleben et al. as 'everything that enables an individual to achieve his or her goals' (Halbesleben 2014, p. 6). Important organizational psychology models, such as the Job Demand–Resources (JD-R) (Demerouti et al. 2001) and the Conservation of Resourses (COR) (Hobfol 1989) models, refer to the concept of 'resources', which is so broad and flexible that it enables an unlimited number of both employee and workplace health determinants. However, it is also criticized for the same reason, i.e. the limited ability to offer a concise meaning of the idea. Arla Day and Karina Nielsen (Day and Nielsen 2017) try to overcome this drawback by defining psychologically healthy workplaces as those workplaces where resources at the individual, group, leader and organizational (IGLO) levels are promoted to ensure employee well-being. The authors thus consider healthy workplaces to be those working environments where, based on the four types of resources, employee health is fostered on the one hand, and high employee performance on the other (Day and Nielsen 2017). Their model offers a framework for classifying workplace resources based on the source of these resources, that is, whether the resources are inherent in the individual, reside within the social context (the work group and the immediate leader) or are provided by the way work is organized, designed and managed, including human resource (HR) practices. The first type of resource proposed in the IGLO model relates to the individual, e.g. the personal characteristics or behaviors that enable the individual to meet the job demands and achieve good results such as high performance, self-esteem and competence, or stress resilience. The second type of resource is at the group level and relates, for example, to social support and/or positive interpersonal relations among staff. Next, the workplace resources at the leader level include leadership characteristics and social interactions between leaders and employees, such as the leadership style and the quality of the leadership–team member exchange. Finally, at the organizational level the resources are associated with the way the work is organized, designed and managed, such as adequate planning of job demands, providing social support, employee growth and development, recognition, and ensuring organizational justice, respect, social inclusiveness and employee involvement.

The research presented in the monograph is focused on workplace resources at all four levels, and their direct and/or indirect associations with employee health and performance. Hildt-Ciupińska, in her study presented in Chapter 2, *Work–Life Balance and Its Determinants among Workers with Dependent Care Responsibilities*, on employees who look after other dependents (child, elderly or person with disabilities, etc.), has analyzed which resources at both individual and organization level have a positive and negative impact on the work–life/life–work balance. Regarding the former relationship, the study results have demonstrated that the main resource at the organizational level constitutes a decent workload management, enabling employees to enjoy annual leave or leisure time and a fine-tuned skills match, contributing to an overall, high job satisfaction level. When these resources are missing, resulting in long working hours and an excessive fatigue, a poor job satisfaction arises, constituting a significant predictor of a negative work–life interaction. At the individual employee level, the determinant of positive work–life/life–work interaction has proved to be a concern for maintaining positive individual health behaviors and effective leisure patterns, which have been found strongly associated with a high work ability. Hence, improved work ability, positive individual health behaviors and high job satisfaction foster a positive work–life and life–work interaction. Achieving work–life balance has been conducive to positive individual health behaviors, as well as increased life and job satisfaction, which are associated with a higher work ability. A novelty of the study has been a finding revealing that a high work ability is yet another individual resource necessary for maintaining a work–home balance.

A study conducted by Mockałło among persons employed in the professional services sector (financial, legal, IT and marketing services) and described in Chapter 3, *Authentic Leadership Style and Worker Innovativeness and Wellbeing: The Role of Climate for Creativity*, has revealed that the authentic leadership style, which is the leadership level resource according to the IGLO model, is both directly and indirectly (by the climate for creativity) associated with the innovativeness of employees. The finding suggests that managers who are authentic leaders promote a climate for creativity and thus contribute to an improved employee performance, i.e. employee innovativeness. Moreover, the results have shown that encouraging employees to develop their full potential and strengths under conditions of trust-based and open relationships, which are a characteristic of the authentic leadership style, has a positive impact on employee work engagement, reduced intention to quit the job and sickness absenteeism. Enterprises that strive to achieve a high performance through employee work engagement, low turnover and absenteeism should therefore ensure that they attract authentic leaders or train their existing leaders in the authentic leadership style. This seems to be particularly important from the perspective of those enterprises whose efficiency heavily relies on the innovativeness and creativity of their employees. The mechanism identified in this research has been based upon the replenishing of healthy workplace resources. Authentic leaders, who are themselves an important resource within the organization, contribute to the growth of the resource pool, fostering a climate for creativity, and thereby unleashing the innovative potential of employees.

The other types of resources have been found to be important from both the employee health (work ability and mental health) and performance (engagement and productivity) points of view in a study conducted by Żołnierczyk-Zreda (Chapter 4, *Selected Employment Characteristics and Employee Health, and Performance: The Mediating Role of Psychological Contract*). One such resource is the psychological contract between the employer and employee, defined as the employee-perceived extent and degree of employer compliance with financial and non-financial obligations. The psychological contract has proved to be a significant mediator in the relationship between the type of employment contract (temporary and permanent), and work ability, mental health, work engagement and employee performance. Permanent-contract employees have been found to be healthier and more productive than those with fixed-term contracts when the psychological contract has been perceived as broad and has not been breached. The occupational status/type of work performed has been identified as a subsequent important resource in terms of employee health and high performance at the individual level. Low-skilled employees performing manual work have been characterized by poorer health and performance, resulting from more frequent employer breaches of psychological contract, compared to high-skilled employees performing non-manual work. Poor individual resources associated with a low occupational status, including a potentially lower ability to negotiate the psychological contract with an employer, may have an adverse impact on employee health. However, they equally affect the employer, significantly reducing the performance of such employees. A novelty of this study has been the aim to demonstrate that a poor-quality psychological contract between employer and employee, including limited access to competence-raising training, affects both an employee's and an organization's health. Additionally, it may lead to a decline in temporary employees' work ability, reducing their chances of breaking out of precarious employment.

Furthermore, the monograph also presents the effects of ignoring the importance of organizational health at the group and organizational level resource (according to the IGLO model), i.e. when employees are overwhelmed with interpersonal conflicts, such as excessive criticism, discrediting, neglect or even workplace bullying.

In a study carried out by Baka in a group of police officers and presented in Chapter 5, *Why Do Employees Behave Counterproductively? The Role of Social Stressors, Professional Burnout and Job Resources*, the findings have revealed that interpersonal conflicts and workplace bullying have a significant, harmful effect on the organization, and have thus been identified as a counterproductive work behavior. Such employee behaviors have been motivated by a desire to retaliate against other employees for bad treatment suffered earlier, which is accompanied by strong emotions of frustration, anger and hostility. Burnout, specifically the decline in employee work engagement, has been found to be a significant mediator of this relationship. Furthermore, the study revealed that even in such difficult psychosocial conditions as interpersonal conflicts and disrespect, some other organizational resources, namely having job control and social support from other coworkers, may effectively alleviate the negative impact of interpersonal conflicts and workplace bullying on the counterproductive work behavior. It thus seems that when these two organizational resources, i.e. increased employee autonomy in performing tasks and decision-making, and a focus on positive interpersonal relations, exist together, they may create an organizational culture that effectively prevents counterproductive work behaviors. The study constitutes one of the very few accounts demonstrating that interpersonal conflicts and workplace bullying lead not only to poor health, particularly the mental health, of employees, but also drastically affect the healthiness of an organization.

Lastly, in the study conducted by Warszewska-Makuch and described in Chapter 6, *Workplace Bullying, Mental Health and Job Satisfaction: The Moderating Role of Individual Coping Styles*, the researcher has examined whether individual employee resources, specifically the task-oriented style of coping with stress, considered as an effective method of diminishing work-related stress, would mitigate the negative effects of workplace bullying on mental health and job satisfaction. The study has also aimed at proving that the other styles, considered in the literature as more dysfunctional, such as the emotion- and avoidance-oriented style, would exacerbate the negative effects of workplace bullying on mental health. The hypotheses have been only confirmed in the group of the so-called *non-victims* of workplace bullying, experiencing this phenomenon very rarely or never. However, in the group of victims of mobbing identified in the study, the above assumptions were not confirmed, as none of the coping styles have turned out to be a significant moderator in the relationship between workplace bullying and mental health. The obtained results suggest that in workplace bullying experiences, none of individual stress management resources are helpful in improving the victim's mental health. Workplace bullying constitutes a highly traumatic experience, which, despite the victim's attempts to curb it or cope with its negative effects, cannot be resolved without a third-party support. It thus seems that individual resources of workplace bullying victims, often replaced by guilt, reduced self-esteem and depression, should be alleviated by resources at other levels, i.e. the group level and leader level, such as social and organizational support, effective anti-mobbing procedures and health recovery strategies.

The presented monograph therefore shows how different workplace resources belonging to four different levels might improve both employee well-being and performance. It has also been shown that these resources can also mediate the influence of other resources or they can alleviate the negative impact of a lack of specific resources on health and performance employees. The overall conclusion for those dealing with interventions and practices focused on the health of employees and organizations is that these interventions should cover all four levels, probably the more, the better (Nielsen et al. 2017).

REFERENCES

Cooper, C., Robertson, I. T., Silvester, J., Burnes, B., Patterson, F., and Arnold, J. 2004. *Work Psychology: Understanding Human Behaviour in the Workplace*. Financial Times Prentice Hall.

Day, A., and K. Nielsen. 2017. What does our organization do to help our well-being? Creating healthy workplaces and workers. In: *An Introduction to Work and Organizational Psychology: An International Perspective*, 3rd edition, eds. N. Chmiel, F. Fraccaroli, and M. Sverke, 295–314. West Sussex: Wiley Blackwell.

Demerouti, E., A. B. Bakker, F. Nachreiner, and W. B. Schaufeli. 2001. The job demands – Resources model of burnout. *J Appl Psychol* 86(3):499–512.

Fitz-enz, J. 1993. The truth about best practices: What they are and how to apply them. *Hum Resour Manage* 36(1):97–103. doi:10.1002/(SICI)1099-050X(199721)36:1<97::AID-HRM16>3.0.CO;2-B.

Grawitch, M. J., M. Gottschalk, and D. C. Munz. 2006. The path to a healthy workplace: A critical review linking healthy workplace practices, employee well-being, and organizational improvements. *Cons Psychol J: Pract Res* 58(3):129–147. doi:10.1037/1065-9293.58.3.129.

Guest, D. 2017. Human resource management and employee well-being: Towards a new analytic framework. *Hum Resour Manag J* 27(1):22–38.

Halbesleben, J. R. B., J. P. Neveu, S. C. Paustian-Underdahl, and M. Westman. 2014. Getting to the "COR": Understanding the role of resources in conservation of resources theory. *J Manag* 40(5):1334–1364.

Hassard, J., T. Cox, S. Leka, and A. Jain. 2017. Healthy organisations: Definitions, models, empirical evidence. *OSH Wiki*. https://oshwiki.eu/wiki/Healthy_organisations:_definitions,_models,_empirical_evidence (accessed September 19, 2019).

Hobfoll, S. E. 1989. Conservation of resources: A new attempt at conceptualizing stress. *Am Psychol* 44(3):513–524.

Jiang, K., D. Lepak, and J. Baer. 2012. How does human resource management influence HRM-organizational performance relationship: A review of quantitative studies. *Int J Manag Rev* 14:391–407.

Nielsen, K., M. B. Nielsen, Ch. Ogbonnaya, M. Känsälä, E. Saari, and K. Isaksson. 2017. Workplace resources to improve both employee well-being and performance: A systematic review and meta-analysis. *Work Stress* 31(2):101–120. doi:10.1080/02678373.2017.1304463.

Schmidt, W. C., L. Welch, and M. G. Wilson. 2000. Individual and organizational activities to build better health. In: *Healthy and Productive Work: An International Perspective*, eds. L. R. Murphy, and C. L. Cooper, 133–147. London: Taylor & Francis.

Van De Voorde, K., J. Paauwe, and M. Van Veldhoven. 2012. Employee well-being and the HRM–organizational performance relationship: A review of quantitative studies. *Int J Manag Rev* 14(4):391–407. doi:10.1111/j.1468-2370.2011.00322.x.

World Health Organization. 2006. *Constitution of the World Health Organization – Basic Documents*, 45th edition, Supplement, October 2006. Geneva: World Health Organization. www.who.int/governance/eb/who_constitution_en.pdf (accessed September 19, 2019).

2 Work–Life Balance and Its Determinants among Workers with Dependent Care Responsibilities

Katarzyna Hildt-Ciupińska

CONTENTS

2.1 INTRODUCTION

Work–life balance (WLB) is defined as an even commitment to professional and private life responsibilities in terms of time allocated and a sense of accomplishment attributed to both commitment areas. There have been identified three distinct components of work–life balance: time (time spent on professional and home duties), commitment (in relation to work and family/social life), and satisfaction (job satisfaction and life satisfaction) (Greenhaus et al. 2003). It should be emphasized that, undoubtedly, work is also a sphere of life; however, the term "life" should be understood in the present context as an area covering such aspects as: family life (childcare and dependent care, e.g. elderly parents, disabled family members), social life (e.g. maintaining social relationships), hobbies, recreation, etc.

Among many human activities, work and family (home) are the two spheres that the greatest attention is devoted to, consuming the most time and energy, while, however, potentially constituting a source of great satisfaction at the same time. The two areas also place different demands on individuals, sometimes contradictory ones, which people strive to equally meet. An ideal result of the interaction between the two spheres is the *work–life balance.*

The demands of various social environments, in particular the work and home (family) environments, create different burdens. According to the *effort-recovery model* developed by Meijman and Mulder (in: Mościcka-Teske and Merecz 2012), for every such burden, an effort is invested in coping with the related demand. The effort-inducing demands trigger many adaptive psychophysiological reactions in the human body (increased heart rate, muscle tension, increased secretion of stress hormones, etc.). The effort-recovery mechanism is a crucial factor enabling adaptation of these reactions. If it is missing or insufficient, a *negative load reaction* develops (Mościcka-Teske and Merecz 2012).

Geurts et al. (2005) applied the effort-recovery model to describe the mutual relationships between work and family life, as well as to develop the *work–home interaction* term, defined as "a process in which the functioning of a person (behavior) in one sphere (e.g. at home) influences (negatively or positively), by the triggered load reactions, the other sphere (e.g. professional life)". In the present paper, the work–life balance has been defined as a positive interaction between work and life (including family and home, in particular), whereas its absence is a negative interaction between these areas.

An absence of work–life balance (negative interaction between work and private life, work–home conflict) may lead to a conflict resulting from the inability to meet the demands of both environments (Geurts et al. 2005). The work–life conflict has an impact on physical and mental health (well-being), perceived quality of life, life and job satisfaction, motivation to work, health behaviors (taking up physical activity, in particular), frequent hazardous behaviors (e.g. excessive alcohol consumption), performance (work performance, in particular), depression, and burnout (Grzywacz 2000; Rose et al. 2007; Hämmig and Bauer 2009). Conflicts are exacerbated by long working hours (overtime), time pressure, and such a work management that impedes the reconciliation* of professional and family roles (Carlson et al. 2000). It should be emphasized that the negative impact of work on home/home on work not only determines the employee's well-being but also the well-being of his/her family.

Family is the most important life value declared by over 80% of Poles (CBOS 2019). Until recently, the work and life spheres have had clear boundaries, owing to the traditional definition of gender roles. At present, these roles have been gradually stirred up due to their complexity, diversity, and time dedicated to fulfilling the traditionally attributed responsibilities. Such a shift has been strongly influenced by globalization processes and the social, economic, and cultural transformation.

Family itself makes up a whole spectrum of various responsibilities. These are primarily childcare and dependent care duties (elderly parents, disabled family members), as well as other reproductive labor activities related to running a household, organizing leisure time, or individual health-related activities, such as regular medical check-ups or health habits (e.g. finding time to practice physical activity).

In the human life cycle, individual life and social roles, as well as the related responsibilities, change over time. For younger persons these obligations are related to caring for small children, whereas ageing adults (50+) support adolescent children, or take care of grandchildren, parents, or chronically ill/disabled relatives.

Ageing workers are often more burdened with caring responsibilities than younger employees. At present, a double burden phenomenon can be increasingly observed in relation to persons aged 50+ who have parallel adolescent children and elderly parents' care responsibilities. This has been referred to as a *sandwich generation* (Evans et al. 2016).

Effective work–life balance policies and practices can have a positive impact on both employers and workers alike, and as a result, can contribute to a sustainable employability, increased motivation and work performance, reduced absenteeism and stress, protection against employees being hired by a competitor, enhanced company brand, and an improved quality of life of workers.

From the worker's point of view, the effective implementation of activities facilitating a reconciliation of work and private life roles can contribute to an improvement of the employee's psychophysical health and general well-being, and boost individual performance, both in the professional sphere and beyond it (Frye and Breaugh 2004; Nylen et al. 2007). There has been a growing awareness among employers of the important role that the work–life balance plays in fostering the well-being of workers and, consequently, the enterprise efficiency. Accordingly, organizational measures, often referred to as "work–life programs", aimed at reconciling work and private life duties have been increasingly introduced by enterprise management. Such schemes are implemented according to workers' needs, and their main objective is to increase the employees' work performance (Moreno-Jiménez et al. 2009).

The most common need reported by workers, and the most simple measure which can be put into practice at the same time, is adequate working time management (Hildt-Ciupińska 2017). This is has been supported by a study conducted in a group of ageing workers (aged 50+), whereby 40% admitted that in their workplaces there

* Reconciling roles, responsibilities helps to maintain balance.

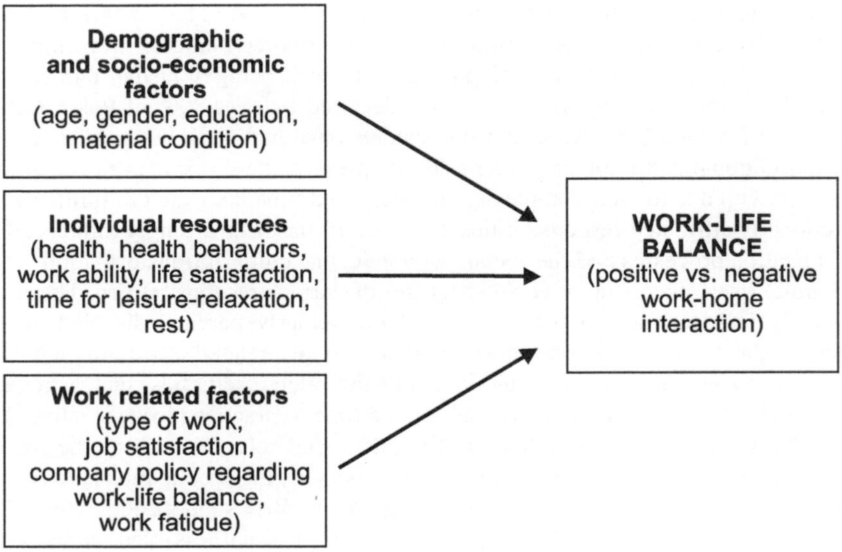

FIGURE 2.1 The hypothetical model of the interdependencies between the study variables.

was a flexible working time schedule (mainly the possibility of starting and ending the working day at hours individually agreed with the supervisor), which was highly appreciated (Hildt-Ciupińska and Bugajska 2013).

The aim of the present study has been to identify the determinants of work–life balance (defined as a positive vs. negative work–home interaction) among workers who have dependent person care responsibilities (childcare, elderly, or disabled family member care, etc.), and relationships between work–life balance and selected study variables.

For the purpose of the study, a hypothetical model of interdependencies between variables has been developed (Figure 2.1), presenting the study variables. The dependent variable has been the work–life balance (positive vs. negative work–home/home–work interaction), whereas independent variables have been demographic factors (age, gender), socio-economic factors, type of work, work ability, health, well-being and health behaviors, job satisfaction and life satisfaction, and organizational measures in terms of reconciling work and life duties (including flexible working time). The assumption has been that the work–life balance, investigated as a positive vs. negative interaction between work and home areas, will be primarily determined by social and economic factors; gender and enterprise measures have been considered important factors in the assessment of the work–life interaction (positive vs. negative).

2.2 METHOD

2.2.1 Participants and Procedure

The study involved 606 workers aged 25 to 65 and over, including 33.0% of respondents aged 25–34 years, 28.1% aged 35–44 years, 19.3% aged 45–54 years, 10.6%

aged 18–24 years, and 8.7% aged 55–64 years. The participants were differentiated by the level of education attained. The highest percentage of the surveyed workers had completed higher education at a master or equivalent level (28.1%), followed by post-secondary or secondary vocational education (27.2%), general secondary education (25.6%), and basic vocational education (17.7%). The study sample was also territorially differentiated – according to the place of residence. The highest percentage of respondents were inhabitants of a city with a population of 50,000 to 100,000 inhabitants (33.5%), followed by rural areas (24.8%), a city with a population up to 50,000 inhabitants (18.0%), and a city with a population of over 500,000 inhabitants (16.2%). The vast majority of the surveyed workers (62.2%) were in a formal relationship, 19.6% in an informal relationship, and 18.0% were single. Also, the majority of respondents had children (81.2%), including children aged up to 2 years (17.2%), children aged 3–5 years (27.7%), children aged 6–12 years (20.5%), children aged 13–16 years (15.2%), and children aged over 17 years (26.1%).

The study participants had both childcare and other dependent care responsibilities. A significant part of respondents (40.3%) took care of a mother or mother-in-law. Every fifth respondent took care of a father or father-in-law, while 18.6% took care of both parents/parents-in-law. Disabled family members were cared for by 14.0% of respondents, other relatives by 10.9%, and other persons by 7.9% of respondents. Considering both children and other dependents, every study participant had care responsibilities for at least one dependent person.

The study sample was selected based on the non-probability quota sampling method, whereby the sampled study participants were determined by industry sector (construction, administration, retail, health care, social services, accommodation and catering), ownership structure (private sector vs. public sector), company size, and gender. The study participants constituted defined quotas of working women (N = 303) and working men (N = 303) evenly distributed among the industry sectors. The study sample (N = 606) was selected out of a population of 100,000 (*confidence level* = 0.95, *estimated fraction size* = 0.5, *maximum error* = 0.04).

2.2.2 Measures

The study survey was conducted using the *paper and pencil interview* (PAPI) technique. It was a direct interview, carried out using a paper questionnaire filled in by a qualified interviewer.

Five standard tools were used in the study. The *Survey Work–Home Interaction* (SWING) questionnaire developed by Nijmegen and Geurts (Polish translation by Mościcka-Teske and Merecz, 2012) was used to assess the work–life balance. It is a positive and negative work–home interaction assessment tool. The questionnaire consists of 22 items covering work–home interaction and comprises 4 distinct subscales: positive impact of work on home (5 items), negative impact of work on home (8 items), positive impact of home on work (5 items), and negative impact of home on work (4 items). The respondent specifies the extent of the impact on a 4-point scale: never (0), sometimes (1), often (2), always (3). An aggregate result is calculated based on the total score obtained in each of the four subscales. Due to the uneven number of items, the authors of the questionnaire (Mościcka-Teske and Merecz 2012) recommend to calculate the mean score per each of the four subscales. (Cronbach's α for four subscales = 0.75–0.86).

The subjective *Work Ability Index* (WAI) developed by the Finnish Institute of Occupational Health (Ilmarinnen 2007) was used to assess the extent to which an employee is capable of doing his or her job. The questionnaire covers seven dimensions measured with a single or several items. The index value is calculated based on the aggregate score obtained for all questionnaire items assessing specific physical and psychosocial job demands, as well as the worker's general health condition and functional capabilities. The WAI possible range of scores is 7–49 points, where 7–27 = poor work ability, 28–36 = moderate work ability, 37–43 = good work ability, and 44–49 = excellent work ability (Cronbach's $\alpha = 0.78$).

Life satisfaction was measured with the *Satisfaction with Life Scale* (SWLS) questionnaire (Diener et al. 1985). It contains five items – statements that the respondent comments on in relation to his/her life. The answers are provided on a 7-point answer scale, where 1 = *do not agree at all*, 4 = *do not agree nor disagree*, and 7 = *totally agree*. The aggregate score is calculated as a sum of each item score and indicates the respondent's overall life satisfaction. The possible range of scores is 5 to 35 points. The higher the score, the greater the life satisfaction.

In interpreting the results one should apply the properties characterizing the sten scale. Results within 1–4 sten are treated as low and 7–10 as high (α-Cronbach: 0.86).

Job satisfaction was measured with a Copenhagen Psychosocial Questionnaire (COPSOQ II), developed by Kristensen et al. (2005), subscale measuring different work aspects (Cronbach's $\alpha = 0.75$).

Individual health behaviors were measured with the Positive Health Behavior Scale (Woynarowska-Sołdan and Węziak-Białowolska 2012), which comprises a set of 33 items grouped in 5 subscales: nutrition, body care (e.g. regular health check-ups), safety (at work, at home, on the road), psychosocial health (sleep, rest, social life), and physical activity. The respondent specifies his/her behavior in relation to each subscale item by providing an answer ranging from almost always (3 points) to almost never (0 points). The scale is used to assess individual health behaviors, and all item scores are summed up to obtain the aggregate score, indicating a low, medium, and high level of individual health behaviors. The reliability of the Positive Health Behavior Scale and all its subscales has been verified using the Cronbach's alpha coefficient. The reliability of the entire scale has been Cronbach's $\alpha = 0.933$. For the whole scale, Cronbach's alpha ranges 0.933; subscale: nutrition, physical activity, safety, sleep and mental health, body care, Cronbach's alpha ranges from 0.65 to 0.95.

In addition, a unique self-assessment questionnaire has been developed for the purpose of the study. The tool consists of four parts:

- Part I. Information on the respondent (age, gender, education, place of residence, marital status, children and other dependent persons, material condition).
- Part II. Information on the company – items related to industry sector (construction, administration, retail, health care and social services, and accommodation and catering), sector (public, private), type of work (intellectual,

physical, mixed), type of contract (full-time vs. part-time, fixed-term vs. permanent), position (management vs. other).

- Part III. Health – items related to self-assessed health status (my health is: very good, good, medium, bad, very bad), ailments experienced in the last six months (ailment frequency – almost every day, more than once a week, almost every week, almost every month, rarely or never) such as headache, abdominal pain, back pain, depression, irritability, nervousness, difficulty in falling asleep, dizziness).
- Part IV. Work–life balance – items related to definition of "work–life balance", self-assessed degree of the reconciliation of work and private life duties (on a scale from 0 – *I do not manage to reconcile at all*, to 10 – *I manage to reconcile very well*), sharing of household tasks with a spouse/partner/other person, time devoted to domestic tasks (number of hours), impact of work on private life and vice versa (very large, large, small, no impact, hard to say), factors conducive to a poor work–life balance (answer choice: working hours, overload of work duties, overload of home duties, ambition to advance professional career, lack of social support at work, lack of family support at home), frequency of staying at work "after hours", taking vacation leave, time for entertainment, recreation, organizational measures facilitating the reconciliation of work and private life duties (e.g. flexible working time, partly working from home), absenteeism caused by dependent person care responsibilities, suggestions for an improved reconciliation of work and private life duties, consequences of work–life imbalance (answer choice: stress, health problems, family problems, lack of job/life satisfaction, lack of motivation to work), and an item on how often (almost always, often, sometimes, almost never) the following situations occur: (a) I have come from work too tired to do domestic tasks, (b) it has been difficult for me to fulfil my family responsibilities because of the amount of time I spent at work, (c) I have had difficulty concentrating on my work because of my family life duties.

2.2.3 Statistical Data Analysis

The statistical data analysis was conducted using the SPSS statistical software. The analysis has included descriptive statistics and a stepwise regression analysis.

2.3 RESULTS

2.3.1 Descriptive Statistics

The results concerning positive vs. negative work–home and home–work interactions, measured with the SWING questionnaire among the surveyed workers, indicate that the respondents experienced more intense positive than negative feelings arising from the work–home and home–work interaction. The mean results for the entire sample ranged from 0.78 to 1.4. The lowest results were obtained in the area of the negative impact of home on work (0.78), which translated into a rare experience

TABLE 2.1
Descriptive Statistics (N = 606)

	Mean	N	SD	Median	Minimum	Maximum
SWING						
Negative impact of work on home	1.0398	606	0.57401	1.0000	0.00	2.75
Positive impact of work on home	1.2201	606	0.50772	1.2000	0.00	3.00
Negative impact of home on work	0.7785	606	0.64692	0.7500	0.00	3.00
Positive impact of home on work	1.4195	606	0.54964	1.4000	0.20	3.00
Positive Health Behavior Scale and Subscales						
Positive Health Behavior Scale (0–99)	56.1820	599	13.89769	55.0000	17.00	95.00
Nutrition subscale (0–27 p.)	13.3079	604	5.23870	13.0000	0.00	27.00
Body care subscale (0–24 p.)	14.7193	602	4.90128	15.0000	1.00	24.00
Psychosocial health subscale (0–21 p.)	11.5769	605	3.84072	11.0000	1.00	21.00
Safety behavior subscale (0–15 p.)	11.2046	606	3.12405	12.0000	1.00	15.00
Physical activity subscale (0–12 p.)	5.3377	604	3.01106	5.0000	0.00	12.00
SWLS (5–35 p.)	20.4505	606	5.32265	20.0000	5.00	35.00
Work Ability Index (0–49 p.)	40.32	604	5.871	41.00	17	49

of this kind of interaction. The highest score was obtained in the area of the positive impact of home on work (1.42) (Table 2.1).

2.3.2 The Predictors of the Positive vs. Negative Work–Home Interaction

In the conducted study, the four dimensions of the SWING questionnaire were dependent variables: positive impact of work on home, negative impact of work on home, positive impact of home on work, and negative impact of home on work. The independent variables respectively were self-assessed health status, frequency of ailments (headache, back pain, abdominal pain, nervousness, irritation, sleeping problems, dizziness), self-assessed reconciliation of work and life duties, work ability, individual health behaviors, life satisfaction and job satisfaction, organizational measures aimed at fostering work–life balance among workers, and other variables defined in the unique study questionnaire (cf. Section 2.2.2, Measures).

2.3.3 The Predictors of the Negative Impact of Work on Home in the Entire Study Sample

The results have showed that in the entire study sample the negative impact of work on home is determined by frequent ailments, low job and life satisfaction, fatigue after returning home from work making it difficult to perform household duties, staying at work "after hours", low work ability, poor individual health behaviors, absenteeism due to dependent person care duties, poor satisfaction with individual skills' match at the workplace. The obtained model explains 42.3% of the dependent variable variance (Table 2.2).

2.3.4 The Predictors of the Negative Impact of Work on Home among Women and Men

The negative impact of work on home among women has been found to be related to poor job satisfaction, low work ability, frequent ailments (such as headaches, back pains), low self-esteem, and poor individual health behaviors. Among men, the relationship with life satisfaction has proved significant, while individual health behaviors have not been found significant (Table 2.3).

TABLE 2.2
The Predictors of the Negative Impact of Work on Home in the Entire Study Sample

	B	Standard Error	Beta	t	Significance (p<)
Health status determined by the frequency of ailments (e.g. headaches, abdominal pain, nervousness, irritability)	−0.031	0.004	−0.296	−7.508	0.000
Job satisfaction	−0.005	0.001	−0.183	−4.497	0.000
Work ability	−0.015	0.004	−0.151	−3.526	0.000
Life satisfaction	−0.009	0.004	−0.086	−2.363	0.018
Individual health behaviors	−0.003	0.002	−0.078	−1.988	0.047
Satisfaction with the skills' match at the workplace	−0.162	0.055	−0.169	−2.941	0.004
Frequent absence due to dependent person care duties	−0.111	0.032	−0.208	−3.448	0.001
Staying at work "after hours"	−0.057	0.021	−0.168	−2.665	0.008

TABLE 2.3
The Predictors of the Negative Impact of Work on Home among Women and Men

	B	Standard Error	Beta	t	Significance (p<)
Women					
Job satisfaction	−0.009	0.002	−0.318	−5.709	0.000
Work ability	−0.027	0.006	−0.285	−4.469	0.000
Health status determined by the frequency of ailments	−0.021	0.006	−0.191	−3.356	0.001
Self-assessed health status	0.129	0.048	0.171	2.691	0.008
Individual health behaviors	−0.005	0.002	−0.114	−2.180	0.030
Men					
Health status determined by the frequency of ailments	−0.039	0.005	−0.398	−7.175	0.000
Job satisfaction	−0.004	0.002	−0.132	−2.323	0.021
Life satisfaction	−0.015	0.005	−0.139	−2.720	0.007
Work ability	−0.013	0.006	−0.133	−2.197	0.029

2.3.5 The Predictors of the Positive Impact of Work on Home in the Entire Study Sample

The analyses have showed that among all the surveyed workers, the following are conducive to the positive impact of work on home: high job and life satisfaction,

TABLE 2.4
The Predictors of the Positive Impact of Work on Home in the Entire Study Sample

	B	Standard Error	Beta	t	Significance
Life satisfaction	0.022	0.004	0.232	5.829	0.000
Individual health behaviors	0.007	0.002	0.207	4.868	0.000
Work ability	−0.015	0.004	−0.169	−4.029	0.000
Job satisfaction	0.002	0.001	0.089	2.007	0.045
Vacation leave	0.081	0.025	0.245	3.236	0.001
Life satisfaction	0.017	0.006	0.174	2.643	0.009
Time for entertainment (theatre/cinema/music concert)	−0.084	0.027	−0.211	−3.114	0.002
Satisfaction with the skills' match at the workplace	−0.155	0.057	−0.188	−2.704	0.008
Absence due to dependent person care duties (child, elderly or disabled person)	−0.091	0.031	−0.196	−2.957	0.004
Time for recreation (garden plot, etc.)	0.049	0.023	0.148	2.148	0.033

positive health behaviors, high work ability, taking vacation leave, satisfaction with individual skills' match at the workplace, finding time for entertainment (cinema, theatre) and recreation, and low absenteeism due to dependent person care duties. The obtained model explains 27.3% of the dependent variable variance (Table 2.4).

2.3.6 The Predictors of the Positive Impact of Work on Home among Women and Men

The positive impact of work on home among women has been found to be most strongly associated with a high life satisfaction and positive health behaviors, while among men it has been additionally associated with a high work ability (Table 2.5).

2.3.7 The Predictors of the Negative Impact of Home on Work in the Entire Study Sample

The analyses carried out for the entire study sample have revealed that those who experience a negative impact of home on work simultaneously report frequent

TABLE 2.5
The Study Variables Explaining the Positive Impact of Work on Home among Women and Men

	B	Standard Error	Beta	t	Significance (p<)
Women					
Life satisfaction	0.021	0.005	0.226	3.954	0.000
Individual health behaviors	0.006	0.002	0.155	2.721	0.007
Men					
Individual health behaviors	0.011	0.002	0.293	5.111	0.000
Life satisfaction	0.023	0.005	0.234	4.287	0.000
Work ability	−0.017	0.005	−0.195	−3.424	0.001

TABLE 2.6
The Negative Influence of Home on Work in the Entire Study Sample

	B	Standard Error	Beta	t	Significance (p<)
Health status determined by the frequency of ailments	−0.045	0.004	−0.388	−10.034	0.000
Work ability	−0.016	0.004	−0.150	−3.714	0.000
Individual health behaviors	−0.006	0.002	−0.124	−3.378	0.001
Life satisfaction	−0.013	0.004	−0.105	−2.956	0.003
Organizational measures facilitating the reconciliation of work and private life duties	−0.040	0.017	−0.081	−2.279	0.023
Time for entertainment (theatre/cinema/concert)	−0.084	0.032	−0.147	−2.617	0.010
Job satisfaction	−0.129	0.065	−0.114	−1.977	0.050

ailments, including headaches, abdominal pains, and dizziness, and note low work ability, poor health behaviors, low job satisfaction and life satisfaction, lack of time for entertainment, and lack of/reduced organizational measures facilitating the reconciliation of work and private life duties. The obtained model explains 53% of the dependent variable variance (Table 2.6).

2.3.8 The Predictors of the Negative Impact of Home on Work among Women and Men

The negative impact of home on work among women has been found most strongly determined by health status (frequent ailments such as headache, back pain, etc.), low work ability, poor individual health behaviors, and low job satisfaction. Similar factors have been identified among men, except for life satisfaction instead of job satisfaction (Table 2.7).

2.3.9 The Predictors of the Positive Impact of Home on Work in the Entire Study Sample

The positive impact of home on work among all the respondents has been strengthened by positive individual health behaviors, a high life satisfaction, and lack of excessive fatigue hindering the fulfillment of household duties. The model explains 22% of the dependent variable variance (Table 2.8).

TABLE 2.7
The Predictors of the Negative Impact of Home on Work among Women and Men

	B	Standard Error	Beta	t	Significance (p<)
Women					
Health status determined by the frequency of ailments	−0.034	0.007	−0.285	−5.022	0.000
Work ability	−0.021	0.006	−0.198	−3.367	0.001
Individual health behaviors	−0.007	0.003	−0.138	−2.618	0.009
Job satisfaction	−0.004	0.002	−0.131	−2.410	0.017
Men					
Health status determined by the frequency of ailments	−0.051	0.006	−0.460	−8.560	0.000
Individual health behaviors	−0.005	0.002	−0.112	−2.151	0.032
Life satisfaction	−0.014	0.006	−0.116	−2.359	0.019
Work ability	−0.014	0.006	−0.126	−2.224	0.027

TABLE 2.8
The Predictors of the Positive Impact of Home on Work in the Entire Study Sample

	B	Standard Error	Beta	t	Significance (p<)
Individual health behaviors	0.008	0.002	0.202	4.989	0.000
Life satisfaction	0.015	0.004	0.144	3.562	0.000
No fatigue hindering the fulfillment of household duties	0.106	0.043	0.169	2.496	0.013

2.3.10 The Predictors of the Positive Impact of Home on Work among Women and Men

Among both women and men, positive individual health behaviors have proved significant in terms of the positive impact of home on work. Additionally, a high life satisfaction has also proved significant among women (Table 2.9).

2.4 DISCUSSION

Facilitating an effective reconciliation of work and private life duties for both women and men, conducive to a sustainable work–life balance, is an important factor in increasing the participation in the labor market, particularly in the context of the current demographic problems. Increased opportunities to reconcile work and private life spheres can have an impact on retaining workers of different ages in employment, a faster return to work after a longer break (e.g. maternity leave), higher job satisfaction (Kafestios 2007), but also greater well-being (good physical and mental health), improved quality of life, and thus an enhanced work performance.

TABLE 2.9
The Variables Explaining the Positive Impact of Home on Work among Women and Men

	B	Standard Error	Beta	t	Significance (p<)
Women					
Individual health behaviors	0.009	0.002	0.217	3.817	0.000
Life satisfaction	0.019	0.006	0.188	3.313	0.001
Men					
Individual health behaviors	0.008	0.002	0.187	3.266	0.001

The aim of the present study has been to identify the most important determinants of maintaining a work–life (family, home, private life) balance among workers with dependent person care responsibilities (children, grandchildren, elderly parents, disabled family members). Many variables have been analyzed, measured both with standard tools and with the unique study questionnaire. Contrary to the expectations, few predictors have proved statistically significant; however, several very important factors have been observed from the worker well-being point of view.

The present research has revealed that two groups of factors are crucial in order to maintain a work–life balance, which in the present study has translated into an experience of a positive impact of work on home and home on work, and an absence of a negative, mutual impact of these two spheres. These factors are individual resources and professional life-related factors.

One of the most important determinants has proved to be individual resources related to good health, healthy lifestyle, high work ability, and life satisfaction. The current research has showed that persons experiencing various ailments (including fatigue, pains of various origins) are more likely to experience negative feelings arising from both work–home and home–work interaction. The study has also revealed that persons who maintain positive health behaviors (practice a healthy lifestyle) have a positive perception of the aforementioned interactions. A lack of work–life balance (conflict between these areas) has a negative impact on various aspects of human functioning. There is a link between a severe work–life conflict and physical and mental health, low quality of life, increased stress levels, and a perception of low job control (Bryson et al. 2007), as well as poor physical activity and increased alcohol consumption (Rose et al. 2007).

The obtained results show that positive work–home/home–work interactions have a significant positive relationship with life satisfaction. Maintaining a work–life balance, particularly minimizing the negative impact of work on home, translates into good health, general well-being, and, consequently, a greater quality of life, which has also been confirmed in other studies (Ruževičius and Valiukaite 2017).

Another group of predictors of the work–life balance are factors related to the professional sphere, including job satisfaction, organizational solutions facilitating the reconciliation of work and home duties (availability of several measures, including flexible working time schedule vs. no solutions), preventing an excessive fatigue hindering the fulfillment of household duties after returning home from work, or avoiding staying at work "after hours". A person satisfied with his/her work and recognized by the management produces higher performance levels at work, which also translate into an increased efficiency at the enterprise level. Achieving a balance between the two spheres is tied to both individual resources and company policy (Walga 2018; Azeem and Akhtar 2014).

One of the predicators of the negative impact of work on home has proved to be excessive fatigue after returning home from work, making it difficult to perform domestic tasks. This may be due to excessive workload, but also to "overtime". The time a person spends at work is an important aspect of the work–life balance. Long working hours (overtime) can have a negative impact on worker health, pose a threat to occupational safety, and induce stress (Matthews et al. 2012). In general, people spend more hours in paid work than on private life activities (excluding sleep). The more time we devote to our professional life, the less time we have left to engage in other activities, such as sports, recreation, or medical check-ups, which are yet crucial to the well-being of workers (Hildt-Ciupińska and Bugajska 2011). The amount and quality of leisure time are vital for general well-being, which, apart from health benefits, generates many other profits in terms of an improved work engagement, performance, job satisfaction, and career prospects and indirectly develops sustainable employability.

Although the reconciliation of work and private life should be facilitated equally for women and men, it should be stressed that women constitute a specific group of workers due to the many burdens they face, particularly those associated with caring for dependents (children, elderly/sick or disabled family members). According to the Statistics Poland data (GUS 2019), women in Poland, unlike men, are more burdened with caring responsibilities, particularly in terms of caring for children under 15 years of age and combining these responsibilities with professional work. However, in the present study group no significant relationships between gender and the assessment of the work–home interaction have been observed. Perhaps this is due to the fact that all respondents had a dependent person under their care, and were therefore similarly burdened with caring responsibilities.

Reconciling work and private life duties should be made easier not only for younger workers (parents), who have difficulties in balancing work and small children care responsibilities, although such workers remain the most vulnerable group in this regard (Craig and Mullan 2011). Older workers also often face this challenge when they cannot afford to retire (for economic reasons) and have to reconcile work with caring for elderly parents or disabled family members (Viitanen 2007; Gardiner et al. 2007). This is often accompanied by fatigue and burnout in this employee group (Ahola et al. 2008).

Taking into account the examined aspects of individual health behaviors (positive health behaviors related to nutrition, physical activity, sleep, rest, safety), it can be concluded that strengthening, supporting, or improving the overall well-being of workers can mitigate the negative feelings arising from the work–home interaction. Health promotion at the enterprise level and an organizational culture that supports work–life balance can help reduce the fatigue and exhaustion resulting from a work–home conflict (Nitzsche et al. 2013a). Negative work–home interactions are associated with an increased risk of depression, whereas positive work–home interactions minimize such a risk (Nitzsche et al. 2013b).

In summary, from the perspective of work–life balance, socio-demographic, economic, and organizational factors seem of a lesser importance. The relationship between positive work–home/home–work interaction and good health, positive health behaviors, high life satisfaction, and work ability is more significant. A greater emphasis should therefore be placed on organizational measures such as the promotion of occupational health at the workplace, as they can play an essential role in reducing the negative impact of work on home and vice versa (Grzywacz 2000). Equally, an adequate workload management, particularly working time management, can have a positive impact on worker well-being, which subsequently translates into an improved worker performance, crucial in achieving enterprise objectives (Obiagelli et al. 2015).

2.5 STRENGTHS AND LIMITATIONS

The results revealing the importance of fostering worker well-being, including the physical and mental health of employees, for maintaining a sustainable work–life balance constitute a strength of the present study. This suggests the need to raise the occupational health awareness and promote worker health in the workplace. Such measures benefit both workers (health, work ability, work–life balance) and employers alike (efficiency of the workers and the company as a whole). The second important group of factors determining the work–life balance has proved to be the work-related factors, i.e. job satisfaction, absence of fatigue hindering the fulfilment of household duties, and recognition of the worker's competence and achievements. The large number of variables determining the work–life balance investigated in the current study has equally been an asset. Nevertheless, not all variables have proved significant in this respect.

There are also some limitations to the present research. Among the surveyed workers, no significant relationships between gender and assessment of work–home interactions have been noted. Perhaps this is due to the fact that all respondents had a dependent person under their care and were therefore similarly burdened with responsibilities. Also, contrary to expectations, age has not proved significant either. Therefore, similar studies should be carried out in a larger random sample.

2.6 CONCLUSIONS

In terms of work–life balance and, consequently, worker well-being, individual resources such as positive health behaviors, healthy lifestyles, high work ability, and life satisfaction seem to be the most important. Workplace health promotion and appropriate workload management solutions (particularly, working time management) that provide opportunities to reconcile work and private life duties can mitigate the stress resulting from a disturbed work–life balance and boost job satisfaction levels among workers.

REFERENCES

Ahola, K., T. Honkonen, M. Virtanen, A. Zromaa, and J. Lönnqvist. 2008. Burnout in relation to age in the adult working population. *Journal of Occupational Health* 50(4):362–365. DOI: 10.1539/joh.M8002.

Azeem, S. M., and N. Akhtar. 2014. The influence of work life balance and job satisfaction on organizational commitment of healthcare employees. *International Journal of Human Resources Studies* 4(2):18–24. DOI: 10.5296/ijhrs.v4i2.5667.

Bryson, L., P. Warner-Smith, P. Brown, and L. Fray. 2007. Managing the work-life roller-coaster: Private stress or public health issue? *Social Science and Medicine* 65(6):1142–1153. DOI: 10.1016/j.socscimed.2007.04.027.

Carlson, D. S., K. M. Kacmar, and L. J. Williams. 2000. Construction and initial validation of a multidimensional measure of work-family conflict. *Journal of Vocational Behavior* 56(2):249–276. DOI: 10.1006/jvbe.1999.1713.

CBOS (Public Opinion Research Centre). 2019. Rodzina – Jej znaczenie i rozumienie. (The family – its meaning and understanding). https://cbos.pl/SPISKOM.POL/2019/K_022_19.PDF (accessed September 3, 2019).

Craig, L., and K. Mullan. 2011. How mothers and fathers share childcare: A cross-national time-use comparison. *American Sociological Review* 76(6):834–861. DOI: 10.1177/0003122411427673.

Diener, E., R. A. Emmons, R. J. Larsen, and S. Griffin. 1985. The satisfaction with life scale. *Journal of Personality Assessment* 49(1):71–75.

Evans, K. L., J. Millsteed, J. E. Richmond, M. Falkmer, T. Falkmer, and s. J. Girdler. 2016. Working sandwich generation women utilize strategies within and between roles to achieve role balance. *PLoS One* 11(6):e0157469. DOI: 10.1371/journal.pone.0157469.

Frye, N. K., and J. A. Breaugh. 2004. Family-friendly policies, supervisor support, work-family conflict, family-work conflict, and satisfaction: A test of a conceptual model. *Journal of Business and Psychology* 19(2):197–220. DOI: 10.1007/s10869-004-0548-4.

Geurts, S. A. E., T. W. Taris, M. A. J. Kompier, J. S. E. Dikkers, M. L. M. van Hooff, and U. M. Kinnunen. 2005. Work home interaction from a work psychological perspective: Development and validation of a new questionnaire, the SWING. *Work and Stress* 19(4):319–339. DOI: 10.1080/02678370500410208.

Greenhaus, H. J., M. K. Collins, and D. J. Shaw. 2003. The relation between work-family balance and quality of life. *Journal of Vocational Behavior* 63(3):510–531.

GUS (Central Statistical Office). 2019. Praca a obowiązki rodzinne w 2018 roku (Reconciliation between work and family life in 2018). https://stat.gov.pl/obszary-tematyczne/rynek-pracy/pracujacy-bezrobotni-bierni-zawodowo-wg-bael/praca-a-obowiazki-rodzinne-w-2018-roku,25,3.html (accessed September 3, 2019).

Hämmig, O., and G. Bauer. 2009. Work-life imbalance and mental health among male and female employees in Switzerland. *International Journal of Public Health* 54(2):88–95. DOI: 10.1007/s00038-009-8031-7.

Hildt-Ciupińska, K. 2017. Równowaga praca – Życie widziana oczami pracowników (Balance of work – Life as seen through the eyes of employees). *Bezpieczeństwo Pracy – Nauka i Praktyka* 4(547):20–23.

Hildt-Ciupińska, K., and J. Bugajska. 2011. Rola zachowań prozdrowotnych w promocji zdrowia pracowników (The role of pro-health behaviours in the promotion of workers' health). *Bezpieczeństwo Pracy – Nauka i Praktyka* 9(480):10–13.

Hildt-Ciupińska, K., and J. Bugajska. 2013. Evaluation of activities and needs of older workers in the context of maintainig their employment. *Medical Section Proceedings* 64(3):297–306.

Gardiner, J., M. Stuart, C. Forde, I. Greenwood, R. MacKenzie, and R. Perrett. 2007. Work–life balance and older workers: Employees' perspectives on retirement transitions following redundancy. *The International Journal of Human Resource Management* 18(3):476–489. DOI: 10.1080/09585190601167904.

Grzywacz, J. G. 2000. Work-Family spillover and health during midlife: Is managing conflict everything? *American Journal of Health Promotion: AJHP* 14(4):236–243.

Ilmarinnen, J. 2007. Work ability index. *Occupational Medicine* 57(2):160. DOI: 10.1093/occmed/kqm008.

Kafestios, K. 2007. Work-family conflict and its relationship with job satisfaction and psychological distress: The role of affect at work and gender. *Hellenic Journal of Psychology* 4(1):15–35.

Kristensen, T. S., and Borg, V. 2005. The Copenhagen Psychosocial Questionnaire – A tool for the assessment and improvement of the psychosocial work environment. *Scand J Work Environ Health* 31(6):438–449.

Matthews, R. A., C. A. Swody, and J. L. Barnes-Farrell. 2012. Work hours and work-family conflict: The double-edged sword of involvement in work and family. *Stress and Health: Journal of the International Society for the Investigation of Stress* 28(3):234–247. DOI: 10.1002/smi.1431.

Moreno-Jiménez, B., M. Mayo, A. I. Sanz-Vergel, S. Geurts, A. Rodríguez-Muñoz, and E. Garros. 2009. Effects of work-family conflict on employee's well-being: The moderating role of recovery experiences. *Journal of Occupational Health Psychology* 14(4):427–440. DOI: 10.1037/a0016739.

Mościcka-Teske, A., and D. Merecz. 2012. Polska adaptacja kwestionariusza SWING do diagnozy interakcji praca-dom i dom-praca (Polish adaptation of swing questionnaire (Survey Work-home Interaction – Nijmegen)). *Medical Section Proceedings* 63(3):355–369.

Nitzsche, A., H. Pfaff, J. Jung, and E. Driller. 2013a. Work-life balance culture, work-home interaction, and emotional exhaustion: A structural equation modeling approach. *J Occup Environ Med.* 55(1):67–73. doi: 10.1097/JOM.0b013e31826eefb1.

Nitzsche, A., J. Jung, H. Pfaff, and E. Driller. 2013b. Employees' negative and positive work-home interaction and their association with depressive symptoms. *American Journal of Industrial Medicine* 56(5):590–598. DOI: 10.1002/ajim.22121.

Nylén, L., B. Melin, and L. Laflamme. 2007. Interference between work and outside-work demands relative to health: Unwinding possibilities among full-time and part-time employees. *International Journal of Behavioral Medicine* 14(4):229–236.

Obiageli, O. L., O. C. Uzochukwu, and C. D. Ngozi. 2015. Work life balance and employee performance in selected commercial banks in Lagos state. *European Journal of Research and Reflection in Management Sciences* 3(4):63. https://www.idpublications.org/wp-content/uploads/2015/05/Abstract-WORK-LIFE-BALANCE-AND-EMPLOYEE-PERFORMANCE-IN-SELECTED-COMMERCIAL-BANKS-IN-LAGOS-STATE.pdf (accessed September 3, 2019).

Rose, S., T. Hunt, and B. Ayers. 2007. Adjust the balance: Literature review life cycles and work life balance. http://www.equalworks.co.uk/resources/contentfiles/4912.pdf (accessed September 3, 2019).

Ruževičius, J., and J. Valiukaite. 2017. Quality of life and quality of work life balance: Case astudy of public and private sectors of Lithuania. *Quality – Access to Success* 18(157):77–81.

Viitanen, T. K. 2007. Informal and formal care in Europe. Discussion paper No. 2648. http://ftp.iza.org/dp2648.pdf (accessed September 3, 2019).

Walga, T. K. 2018. Job satisfaction and satisfaction with work-life balance across cultures. *Journal of Intercultural Management* 10(2):159–179. DOI: 10.2478/joim-2018-0013.

Woynarowska-Sołdan, M., and D. Węziak-Białowolska. 2012. Analiza psychometryczna Skali Pozytywnych Zachowań Zdrowotnych dla dorosłych. (Psychometric analysis of Positive Health Behaviours Scale for adults). *Probl Hig Epidemiol* 93(2):369–376.

3 Authentic Leadership Style and Worker Innovativeness and Wellbeing

The Role of Climate for Creativity

Zofia Mockałło

CONTENTS

3.1 INTRODUCTION

The challenges of the current economic system and the new world of work have recently spawned a vast body of research on the predictors of worker capacity for innovation and only a seemingly separate question of worker wellbeing. However, both factors – worker innovativeness and wellbeing – seem to have common sources in psychosocial work characteristics, including organizational leadership behavior and the corresponding organizational climate.

3.1.1 Authentic Leadership Style and Worker Innovativeness and Wellbeing

Previous research has shown that the leadership style presented by management has a significant impact not only on worker performance, but also on the broadly understood worker wellbeing and life in the organization (e.g. Skakon et al. 2010; Donaldson-Feilder et al. 2013). The leadership style is related to sickness absenteeism, or presenteeism (Nyberg et al. 2008), retention, job satisfaction, burnout, alienation (Nyberg et al. 2005), psychological capital, work engagement, positive and negative affect (McMurray et al. 2010), mental wellbeing, and work–life balance (Munir 2012). Over the years, many studies have also shown the links between leadership behavior and occupational stress (Donaldson-Feilder et al. 2013), to the extent that the relations between managers and employees have been considered the most common source of stress declared by workers (Tepper 2000; Donaldson-Feilder et al. 2013). These relationships can be direct, but also indirect, determined by working conditions created by the management, including psychosocial working conditions (Widerszal-Bazyl 2003; Nyberg et al. 2005; Purkiss and Rossi 2008).

Similar relationships have been observed in studies on leadership behaviors and innovation/innovativeness. Sound leadership behaviors seem to foster creativity and innovativeness among workers (e.g. Herrmann and Felfe 2014; Matzler 2008). An appropriate leadership style is also related to clear team objectives, high levels of participation, commitment to excellence, and support for innovation (West et al. 2003) – factors that seem to be directly related to worker innovativeness, also through the empowerment of employees (Burpitt and Bigoness 1997). Researchers have also proven that leadership is related to organizational climate and its dimensions, such as autonomy, cohesion, trust, pressure, support, recognition, fairness, and supervisor encouragement of innovation (McMurray et al. 2010), which also seem to be related to both worker innovativeness and wellbeing.

The relationships between leadership and innovativeness are mediated and moderated by a number of factors at the individual level, e.g. creative self-efficacy, organizational-based self-esteem, and self-presentation (Denti and Hemlin 2012), the group level, such as team process (West et al. 2003), cooperation (Paulsen et al. 2009), team reflection, team heterogeneity, and task characteristics (Denti and Hemlin 2012), or the organizational level, including support for innovation (Gumusluoğlu and Ilsev 2009), organizational structure, and lastly organizational culture (Denti and Hemlin 2012).

In view of the multitude of leadership styles analyzed, the research has been increasingly focused on exploring not only the question of the leadership behaviors that nurture greater worker performance and creativity, but also the type of personal attributes endorsed by the leader who presents such behaviors. Among many ideas, the concept of authentic leadership, developed by Avolio et al. (2004), has attracted attention in recent years. According to the authors, transparent work conduct based upon being true to personal ethics and beliefs, and the courage to be a true self at the workplace have begun to pose a real challenge. The leaders who succeed in doing so are characterized by transparency, trust, honesty, and high moral standards. Such persons can be described as authentic leaders who not only are honest with themselves but also enable their coworkers to embrace a transparent and fair work attitude (Gardner et al. 2005).

The authentic leadership style includes the following four components:

1. Self-awareness
2. Transparency (relational transparency)
3. Balanced processing
4. Ethical/moral (internalized moral perspective)

Genuine and transparent interpersonal relationships fostered by the authentic leadership style, coupled with personal development opportunities and psychological comfort at the workplace, produce high levels of employee work engagement. Moreover, authentic leaders motivate workers to unveil their potential and facilitate a more aligned person–job fit, further increasing employee work engagement (Gardner et al. 2005).

According to the authors, the authentic leadership style should also determine worker wellbeing in such aspects as vitality, life satisfaction, and mental health (Ryan 2001; in: Gardner et al. 2005). International research has yielded mixed results. For example, Kim (2018) proved that authentic leadership was related to eudemonic wellbeing, and not to hedonic wellbeing (life satisfaction). It was also showed that authentic leadership had an influence on psychological wellbeing at work, indirectly, through the work climate (Nelson et al. 2014). In turn, Rahimnia and Sharifirad (2015) showed that authentic leadership had a positive impact on job satisfaction and only an indirect impact on perceived stress and stress symptoms in followers. Laschinger and Fida (2014) showed that authentic leadership was related to a greater job satisfaction and a lower level of professional burnout and mental health problems among workers.

It has been assumed that the authentic leadership style, by emphasizing talent discovery and employee development, should be related with worker innovativeness. By virtue of being an authentic leader, the manager becomes a positive role model for employees and shapes organizational ethics based on social support and a genuine interest in worker wellbeing, including the development of individual strengths (Avolio et al. 2004). The authentic leadership style fosters the strengths of both leaders and employees alike. By encouraging the latter to work on their talents, leaders stimulate employee work engagement, inciting workers to think outside of the box and go beyond the usual schemes and basic tasks, which is a vital component of innovation. Empirical research has produced inconsistent results. Direct associations of the authentic leadership and employees innovativeness have been shown by Černe, Jaklič, and Škerlavaj (2013), among others. On the other hand, Elrehail et al. (2018) have not proven any significant relationships between such a leadership style and innovativeness. Zhou, Ma, Cheng, and Xia (2014), investigating the mediators of this relationship, revealed the mediating role of positive emotions, but not the negative ones. As mentioned earlier, the authentic leadership model assumes the mediating role of the organizational climate. However, the organizational climate in the authentic leadership model has been described in rather general terms so far. Hence, the present research focuses on an aspect of the organizational climate, related to the strength development orientation, i.e. the climate for creativity.

3.1.2 Climate for Creativity

Social development and economic progress are conditioned by a broadly defined social culture, i.e. values, attitudes, beliefs, and the myriad of social and interpersonal relations. Similarly, organizational culture is shaped by social and economic development patterns occurring in the enterprise. The shared, core enterprise values at the behavioral level create the organizational climate, i.e. "permanent patterns of behaviors, attitudes and feelings that characterize life in the organization" (Ekvall 1996). As such, the organizational climate is consistent with the creativity and innovativeness of workers. A work atmosphere conducive to innovation and change has been defined as a climate for creativity or a climate for innovation (Ekvall 1996; West et al. 2002; Amabile and Gryskiewicz 1989). Such a climate fosters the development of new products and solutions, as well as progress and the uptake of novel ideas.

A climate for creativity is defined as an atmosphere that nurtures creative activities, with a particular emphasis on stimulators (influencing and stimulating creative activity) and inhibitors (factors constraining the development of creativity) present in the environment (Karwowski 2009b). The climate for creativity is critical given the importance of enterprise performance and efficiency. In the absence of such a climate, there is no room for innovative thought, and thus no patented or novel solutions are produced, which may translate into adverse outcomes for the enterprise development (or lack thereof) and economic loss (Nęcka 2001), also to the detriment of employee personal development, work engagement, and physical and mental wellbeing.

The literature provides various meanings for the concept of a climate for creativity; however, some common characteristics can be identified. These are i.a.: social support from coworkers, positive relations with managers, organizational resources, challenging opportunities, mission clarity, positive interpersonal exchange, intellectual stimulation, supervisor support, reward orientation, flexibility and risk-taking, product orientation, participation, and organizational integration (Karwowski 2009a).

A leading definition of the climate for creativity is Goran Ekvall's concept (1996) introducing factors (organizational characteristics), contributing to the organizational climate that fosters creativity and innovation. These are mission and strategy; structure and size; leadership behaviors; organizational culture; resources and technology; requirements and tasks; individual abilities; management practices; systems, policies, and procedures; and needs, motivation, and individual working styles.

Whereas the climate for creativity itself consists of the following organizational aspects: challenge, freedom, trust and openness, idea time, playfulness and humor, idea support, debate, risk-taking, dynamism, and conflict.

Among these factors, only the conflict at work is negatively correlated with creativity: the more interpersonal conflicts are present, the lesser the likelihood of innovative thought. The other factors discussed should have a positive impact on creativity and innovation (Ekvall 1996).

The model developed by Ekvall (1996) introduces the climate for creativity as a mediating factor in the relationship between job resources and work outcomes and worker wellbeing. The climate for creativity has a significant impact on organizational processes such as problem-solving, decision-making, communication, coordination, and control, as well as psychological processes such as learning, creativity, motivation, or loyalty to the organization. In turn, work outcomes mutually affect both job resources and the organizational climate.

The model has been supported by Ekvall in his study among academic teachers. The job resources and organizational climate were found to be strongly related to the organizational creativity, whereby it further stimulated leadership behaviors, which has proven that the organizational climate, creativity, and leadership style are intertwined (Ekvall and Ryhammar 1999).

Interestingly, two of the surveyed leadership characteristics were strongly correlated with the climate for creativity: the change/development-orientation and the relationships/employee-orientation. However, the leadership characteristics were not directly related to innovative outcomes as this relationship was mediated by the climate for creativity. In turn, such conditions, i.e. the organizational climate conducive to creativity, also had an impact on leaders, strengthening those behaviors that fostered the capacity for innovation. If managers had shown concern for social relations at the workplace, an atmosphere of trust and support would have been created and would have encouraged the management to foster a positive work culture. Alternatively, the management could have adapted the team management style to the organization's climate for creativity (Ekvall and Ryhammar 1999).

Many studies have reported on the relationship between the organizational climate for creativity and the enterprise and worker innovativeness (e.g. Ekvall 1996; Ekvall and Ryhammar 1999; Barrett et al. 2005; Ismail 2005; Sellgren et al. 2008). However, empirical studies verifying theoretical concepts of organizational climate for creativity have yielded inconsistent results. The differences can be identified in these organizational climate aspects that are related to enterprise performance.

While some results have shown that the climate for creativity is determined by the following factors: challenge, freedom, support, lack of fear, space for an open discussion, and moderate risk-taking (Nęcka 2001), other research has proven that only challenge and the possibility to discuss in a relaxed atmosphere are relevant to the climate for creativity (Ismail 2005).

A comprehensive meta-analysis was carried out by Hunter, Bedell, and Mumford (2007), who analyzed the relationship between different aspects of organizational climate and indicators of worker innovative behavior. The analyses showed that climate for creativity was strongly significantly related to creative work outcomes, regardless of the study methodology, population, or moderating factors. Challenge, intellectual stimulation, and positive interpersonal exchange were most strongly related to innovative work outcomes. The researchers also highlighted the importance of such factors as social support, resources, and autonomy that the management could shape. Hence, the supervisor's role in stimulating the creativity and innovativeness of workers by the fostering of a positive work culture cannot be underestimated (Hunter et al. 2007).

The relationships between leadership style and climate for creativity and worker innovativeness were examined i.a. by Ekvall (1996), Sellgren, Ekvall, and Tomson (2008), and Aarons and Sommerfeld (2012). The research results supported the strong relationship between leadership style and climate for creativity, the latter providing space for the leadership style to shape enterprise innovation and efficiency outcomes. However, there have been very few studies analyzing the conditions generating such a relationship (Jung 2001).

Moreover, it is worth noting that the two models mentioned, the climate for creativity model and the authentic leadership model, are yet to be empirically verified. Hence, there seems to be a need for such research, particularly considering the paucity of similar studies in Poland.

3.1.3 The Present Study

The aim of the present study has been to examine whether authentic leadership is a predictor of worker innovativeness and wellbeing. The aim has also been to investigate whether the climate for creativity plays a mediating role in these relationships.

3.1.4 Study Hypotheses

H1: The authentic leadership style is positively related to worker innovativeness.
H2: The authentic leadership style is positively related to worker wellbeing.
- H2a: The authentic leadership style is positively related to employee work engagement.
- H2b: The authentic leadership style is negatively related to stress.
- H2c: The authentic leadership style is negatively related to worker intent to quit.
- H2d: The authentic leadership style is negatively related to worker sickness absence.

H3: Climate for creativity is positively related to worker innovativeness.

H3a: Challenge is positively related to worker innovativeness.
H3b: Freedom and debate are positively related to worker innovativeness.
H3c: Conflict is negatively related to worker innovativeness.

H4: Climate for creativity is positively related to worker wellbeing.
H4a: Challenge is positively related to employee work engagement.
H4b: Challenge is negatively related to stress.
H4c: Challenge is negatively related to worker intent to quit.
H4d: Challenge is negatively related to worker sickness absence.
H4e: Freedom and debate are positively related to employee work engagement.
H4f: Freedom and debate are negatively related to stress.
H4g: Freedom and debate are negatively related to worker intent to quit.
H4h: Freedom and debate are negatively related to worker sickness absence.
H4i: Conflict is negatively related to employee work engagement.
H4j: Conflict is positively related to stress.
H4k: Conflict is positively related to worker intent to quit.
H4l: Conflict is positively related to worker sickness absence.

H5: Climate for creativity is a mediator in the relationship between authentic leadership and worker innovativeness and wellbeing.
H5a: Challenge is a mediator in the relationship between authentic leadership and worker innovativeness and wellbeing.
H5b: Freedom and debate is a mediator in the relationship between authentic leadership and worker innovativeness and wellbeing.
H5c: Conflict is a mediator in the relationship between authentic leadership and worker innovativeness and wellbeing.

3.2 METHOD

3.2.1 Participants and Procedure

The study was conducted in a group of 270 employees working in professional services sector enterprises, i.e. companies focused primarily on the provision of business consultancy services, offering specialist knowledge and methods to execute a specific type of business contracts.

The study group employees were selected from the following types of enterprises: financial and accounting services, graphic design, legal services, IT services, real estate and facility management agency, and research and marketing services.

The mean age in the study group was 37.55 years (*SD* = 11.62). The youngest respondent was 19 years old, the oldest was 72 years old. The study group consisted of 142 women and 116 men (53% and 43% respectively). The average work experience in the study group was 12.6 years (*SD* = 11.6). The shortest professional career was 2 months, the longest 54 years. The majority of respondents had a higher education degree (172 employees), and 70 respondents were educated up to the secondary education level. The survey was conducted as a cross-sectional study. The respondents filled in the questionnaires using the paper and pencil interview (PAPI) method in the presence of an interviewer who collected the completed questionnaires and placed them in envelopes.

3.2.2 Measures

3.2.2.1 Authentic Leadership Style

The *Authentic Leadership Questionnaire* (*ALQ*) (Avolio et al. 2007) was used to measure the leadership style. The *Authentic Leadership Questionnaire* measures the following four dimensions of authentic leadership:

1. Self-awareness (four items) – defines the extent to which a leader is aware of his or her strengths and limitations, as well as the way others perceive him/her and the influence he or she has on them (e.g. "My leader seeks feedback to improve interactions with others").
2. Transparency (five items) – defines the extent to which the leader is open to others and thus honestly expresses his/her opinions, presents true thoughts and emotions, and honestly admits to the mistakes made (e.g. "My leader says exactly what he or she means").
3. Ethical/moral (four items) – determines whether the leader sets high moral and ethical standards for behavior in the company (e.g. "My leader makes difficult decisions based on high standards of ethical conduct").
4. Balanced processing (three items) – determines the extent to which the leader seeks feedback from employees before making important decisions (e.g. "Listens carefully to different points of view before drawing conclusions").

The questionnaire comprises 16 items which all describe the different behaviors and actions of a leader related to the authentic leadership style. In the employee version, respondents are asked to indicate on a four-point scale the frequency with which the following statements match the management style represented by their supervisor. The questionnaire's reliability is high, Cronbach's $\alpha = 0.92$.

3.2.2.2 Climate for Creativity

The *Workplace Creative Climate Questionnaire* (Karwowski 2009a) was used to measure the organizational climate for creativity. The questionnaire comprises 48 items that measure the 3 basic dimensions of the climate for creativity:

1. Freedom and debate (23 items) – this factor describes a working atmosphere that fosters employee confidence in autonomy on the one hand, and participation in decision-making on the other. The scale's reliability is Cronbach's $\alpha = 0.94$.

   This factor consists of three sub-scale factors:
   - Freedom (14 items): e.g. "I think that the strength of my company is that employees are responsible for the decisions they make".
   - Trust (six items): e.g. "I trust most of my employees".
   - Calm (three items): e.g. "Company works at a balanced pace – haste will not improve the results, but will spoil the atmosphere".
2. Challenge (17 items) – describes such an enterprise culture where employees meet difficult and complicated problems, and take up the

challenge of finding the solution; a workplace where there is an element of uncertainty and risk, however, this is a constructive risk (e.g. "I have repeatedly performed very uncertain tasks"). The scale's reliability is Cronbach's $\alpha = 0.82$.

3. Conflict (eight items) – this factor corresponds to interpersonal conflicts that occur among employees, as well as a stressful and uncertain atmosphere that generates and nurtures these conflicts (e.g. "Sometimes we (my colleagues and I) failed to achieve our goals due to quarrels and misunderstandings"). The reliability of this scale is not high, Cronbach's $\alpha = 0.60$, and as the author of the tool declares –

> this may be partly due to a few items, however, given the importance of this factor for the understanding of organizational climate for creativity and its determinants, a decision was taken to further analyze and refine this scale in subsequent versions of the tool.

3.2.2.3 Innovativeness

Innovativeness was measured using the *Innovative Output* scale (De Jong and Den Hartog 2010), which consists of six items that correspond to the frequency of proposing and implementing activities related to new products, services, working methods, knowledge, and the market (customer groups). An example of a question is: "In your job, how often do you make suggestions to improve current products or services?" Respondents are asked to provide the answer on a five-point scale (1 – never; 5 – always). The scale reliability is sufficient, Cronbach's $\alpha = 0.82$.

3.2.2.4 Worker Wellbeing

Worker wellbeing was measured using several indicators: work engagement as an indicator of eudaimonic wellbeing, stress symptoms and intent to quit as indicators of poor wellbeing, and sickness absence as a more objective indicator of general wellbeing.

1. Work engagement – measured using the abbreviated version of the UWES questionnaire (Schaufeli et al. 2004). A three-item scale was used in the study, which had already been used in Finnish (Kauppinen et al. 2007) and Polish (Wiezer et al. 2011) studies. Recently, this version has been published in Schaufeli et al. (2017) as the ultra-short measure of work engagement, UWES-3 (e.g. "At my work, I feel bursting with energy"). The respondents were asked to provide the answer by ticking the appropriate value on a scale from 0 to 6, where 0 means "never" and 6 means "always/every day." The scale reliability is sufficient, Cronbach's $\alpha = 0.84$.
2. Work-related stress was assessed using a single-item measure of stress symptoms (Elo et al. 2003), modified by adding the wording "work-related" (Wiezer et al. 2011). The question was: "Stress means a situation in which a person feels tense, restless, nervous or anxious or is unable to sleep at night because his/her mind is troubled all the time. Do you feel this kind of

TABLE 3.1
Descriptive Statistics of the Study Variables ($N = 270$)

Variable	*N*	*M*	*SD*	*Min.*	*Max.*
Freedom and debate	267	3.43	0.62	1.35	4.91
Challenge	267	3.62	0.47	1.71	4.88
Conflict	267	2.46	0.65	1	4.63
Authentic leadership (total score)	266	2.68	0.67	0.25	4
Innovativeness	266	3.54	0.59	2	5
Work engagement	266	5.69	1.06	1.67	7
Sickness absence	256	4.23	9.03	0	100
Intent to quit	266	2.31	1.07	1	5
Work-related stress	264	3.88	1.8	1	7

$^{*}p < 0.05$, $^{**}p < 0.01$. $^{***}p < 0.001$.

work-related stress these days?" The response was recorded on a five-point Likert scale varying from 1 "not at all" to 5 "very much".

3. Intent to quit – the variable was measured with the question: "Do you plan to stay in your current job for the next five years?" Respondents were asked to indicate the answer on a scale from 1 to 5, where 1 means "definitely yes" and 5 means "definitely no".
4. Sickness absence – the number of days spent on sick leave over the last year.

3.2.3 Statistical Analysis

In order to test the mediating role of climate for creativity in the relationship between authentic leadership and worker innovativeness and wellbeing, a series mediation analysis with three mediators (*challenge*, *conflict*, *freedom and debate*) was conducted using the SPSS 23 Macro Process with bootstrapping (5,000 bootstrap samples; Model 4, Hayes 2017).

3.3 RESULTS

3.3.1 Descriptive Statistics

The descriptive statistics analysis of the study variables shows that the surveyed workers assessed their work engagement and innovativeness quite strongly (Table 3.1). Sickness absence in the surveyed group was around four days on average. The mean response values indicate that the respondents worked in organizations characterized by a climate for creativity, where supervisors displayed certain authentic leadership behaviors; however, the respondents also experienced symptoms of stress.

3.3.2 Mediation Analysis

3.3.2.1 The Relationship between the Authentic Leadership Style and Worker Innovativeness, and a Mediating Role of Climate for Creativity

The tested model had a good data fit, $F(4.260) = 20.03$; $p < 0.001$ and explained 24% of the dependent variable variance.

The independent variable (authentic leadership) was significantly related to two out of the three climate for creativity aspects analyzed, i.e. the *challenge*, and the *freedom and debate*. *Conflict*, although significantly positively related to worker innovativeness, did not have a mediating role in the relationship between the independent and dependent variables (Figure 3.1.).

The analysis of indirect effects (Table 3.2) proved that the relationship between authentic leadership and worker innovativeness was partly mediated by two climate for creativity aspects. The higher the level of authentic leadership, the higher the level of *challenge* and, as a result, the higher the level of worker innovativeness. Similarly, the higher the level of the authentic leadership style, the higher the level of *freedom and debate*, which, however, was conducive to lower levels of worker innovativeness.

3.3.2.2 The Relationship between the Authentic Leadership Style and Worker Wellbeing, and the Mediating Role of Climate for Creativity

3.3.2.2.1 Work Engagement

The tested model fit the data, $F(4.260) = 9.34$; $p < 0.001$ and explained 13% of the dependent variable variance (Figure 3.2).

The independent variable (authentic leadership) was significantly related to two out of the three climate for creativity aspects analyzed, i.e. the *challenge* and the

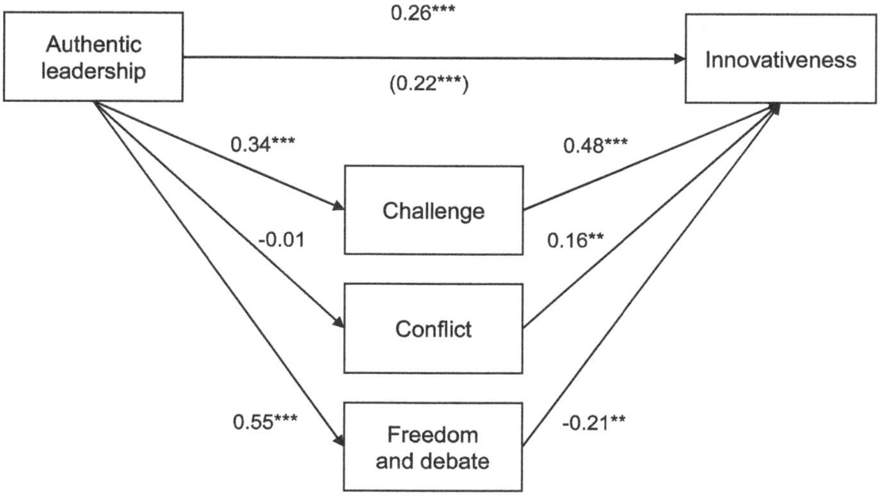

FIGURE 3.1 Standardized regression coefficients for the relationship between authentic leadership and worker innovativeness as mediated by climate for creativity (challenge, conflict, freedom and debate). ***$p < 0.001$. **$p < 0.01$. *$p < 0.05$.

TABLE 3.2
Estimation of Indirect Climate for Creativity Effects in the Relationship between Authentic Leadership and Worker Innovativeness (N = 270)

	Indirect Effects B (BootSE)	95% Boot CI	
		BootLLCI	BootULCI
Total	0.04 (0.05)	–0.06	0.14
Authentic leadership -> challenge -> worker innovativeness	0.16 (0.04)	0.09	0.24
Authentic leadership -> conflict -> worker innovativeness	–0.002 (0.01)	–0.03	0.02
Authentic leadership -> freedom and debate -> worker innovativeness	–0.12 (0.05)	–0.22	–0.02

Note: B = standardized indirect effect; BootSE = bootstrapped standard error; Boot CI = bootstrapped confidence interval; BootLLCI = bootstrapped lower limit of the confidence interval; BootULCI = bootstrapped upper limit of the confidence interval; level of confidence for the confidence intervals = 95%.
***$p < 0.001$. **$p < 0.01$. *$p < 0.05$.

freedom and debate, whereas the association with the third climate for creativity aspect was found insignificant. Given that the *conflict* was the only climate for creativity aspect significantly related to work engagement (dependent variable), it can be concluded that the climate for creativity did not play a mediating role in the relationship between authentic leadership and work engagement. Based on the obtained results, the work engagement level was explained by the authentic leadership style and the *conflict*: the greater the authentic leadership a manager displayed, the more

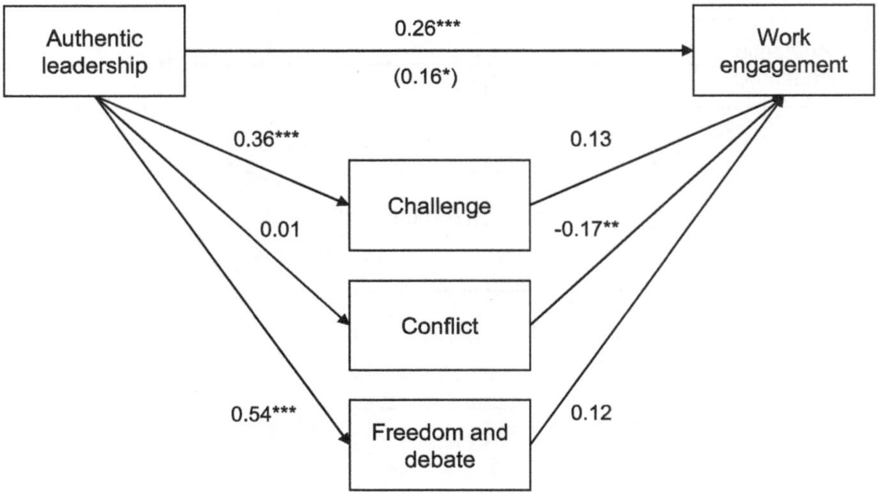

FIGURE 3.2 Standardized regression coefficients for the relationship between authentic leadership and employee work engagement as mediated by climate for creativity (challenge, conflict, freedom and debate). ***$p < 0.001$. **$p < 0.01$. *$p < 0.05$.

engaged at work employees were. In turn, the greater the interpersonal conflicts in the workplace, the weaker the employee work engagement.

3.3.2.2.2 Stress

The tested model had a good data fit, $F(4.258) = 9.05$; $p < 0.001$ and explained 12% of the dependent variable variance. The authentic leadership was positively related to the *challenge* as well as the *freedom and debate*. All the climate for creativity aspects were found significantly related to the stress experienced by workers: the greater the *challenge* and the *conflict*, the higher the level of work-related stress. In turn, the greater the *freedom and debate*, the lower the work-related stress among the workers (Figure 3.3).

The relationship between the independent variable and the dependent variable proved to be insignificant both before and after the introduction of the mediating factors into the model. However, according to a recent mediation analysis approach (Shrout and Bolger 2002; Hayes 2009; Hayes 2017), the mediational effect can be considered significant even if the total effect (c) is insignificant. In order to verify the significance of the mediational effect, an analysis of indirect effects should be carried out (Table 3.3).

The indirect effects' analysis showed that there were two mediational effects corresponding to the *challenge* and the *freedom and debate*. The more authentic the leadership style adopted by the management, the higher the level of *challenge* faced by employees, resulting in increased stress levels among the workers. In turn, the greater the authentic leadership behaviors displayed by managers, the higher the level of *freedom and debate* perceived by employees, which led to a reduction in the level of the work-related stress experienced by the workers.

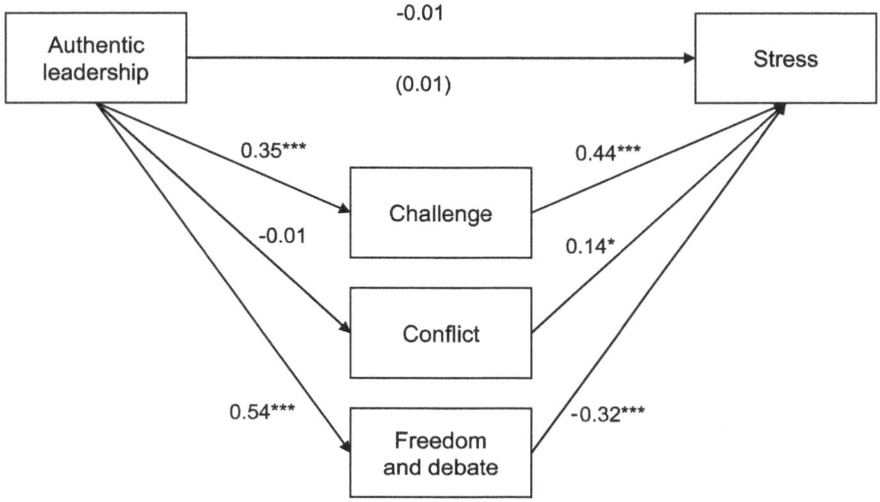

FIGURE 3.3 Standardized regression coefficients for the relationship between authentic leadership and work-related stress experienced by the workers as mediated by climate for creativity (challenge, conflict, freedom and debate). ***$p < 0.001$. **$p < 0.01$. *$p < 0.05$

TABLE 3.3
Estimation of Indirect Climate for Creativity Effects in the Relationship between Authentic Leadership and Work-Related Stress Experienced by the Workers (*N* = 270)

	Indirect Effects B (BootSE)	95% Boot CI	
		BootLLCI	BootULCI
Total	–0.02 (0.05)	–0.12	0.08
Authentic leadership -> challenge -> stress	0.15 (0.04)	0.08	0.24
Authentic leadership -> conflict -> stress	–0.001 (0.01)	–0.02	0.02
Authentic leadership -> freedom and debate -> stress	–0.17 (0.05)	–0.28	–0.07

Note: B = standardized indirect effect; BootSE = bootstrapped standard error; Boot CI = bootstrapped confidence interval; BootLLCI = bootstrapped lower limit of the confidence interval; BootULCI = bootstrapped upper limit of the confidence interval; level of confidence for the confidence intervals = 95%.

3.3.2.2.3 Intent to Quit

The tested model had a good data fit, $F(4.260) = 4.87$; $p < 0.001$ and explained 7% of the dependent variable variance. The leadership style was negatively related to the intention to quit the job: the higher the level of authentic leadership among the managers, the lower the intent of the workers to quit the job. The authentic leadership was positively related to the *challenge* as well as the *freedom and debate*. The relationship between the independent and dependent variables was slightly mitigated when the mediators were introduced into the model. However, out of the three climate for creativity aspects, only the *conflict* was significantly related to the intention to quit the job: the higher the level of interpersonal conflict, the greater the intention of workers to quit the job. This means that no indirect effects have been observed in this model and, therefore, the hypothesis of a mediating role of climate for creativity has not been confirmed (Figure 3.4).

3.3.2.2.4 Sickness Absence

The tested model fit the data, $F(4.250) = 3.86$; $p < 0.01$ and explained the 6% of the dependent variable variance. The authentic leadership was positively related to the *challenge*, and the *freedom and debate*, and there was no significant relationship observed between authentic leadership and *conflict*. Two out of the three climate for creativity aspects were found related to sickness absence: the *challenge* (negative association) and the *freedom and debate* (positive association).

As shown in Figure 3.5, the relationship between the independent variable and dependent variable was reinforced after the introduction of the mediating factors into the model (before the introduction of the mediating factors, the association had been at the statistical tendency level) which could suggest a mechanism of suppression.

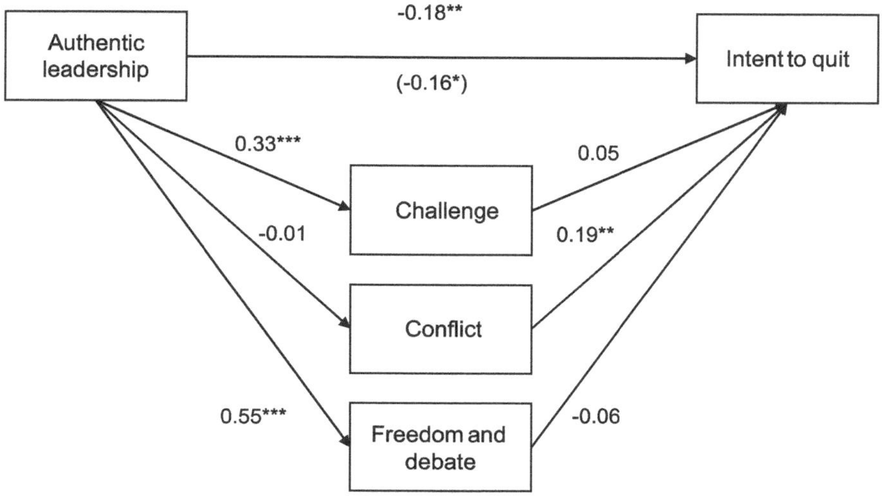

FIGURE 3.4 Standardized regression coefficients for the relationship between authentic leadership and worker intent to quit as mediated by climate for creativity (challenge, conflict, freedom and debate). ***$p < 0.001$. **$p < 0.01$. *$p < 0.05$.

As shown in Table 3.4, with regard to one mediator, i.e. the *challenge*, a significant indirect effect was observed. The higher the level of authentic leadership displayed by the managers, the greater the *challenge* perceived by the employees, which translated into a reduced sickness absence. However, given that the upper limit of the confidence interval is very close to the zero value, the result should be treated with caution.

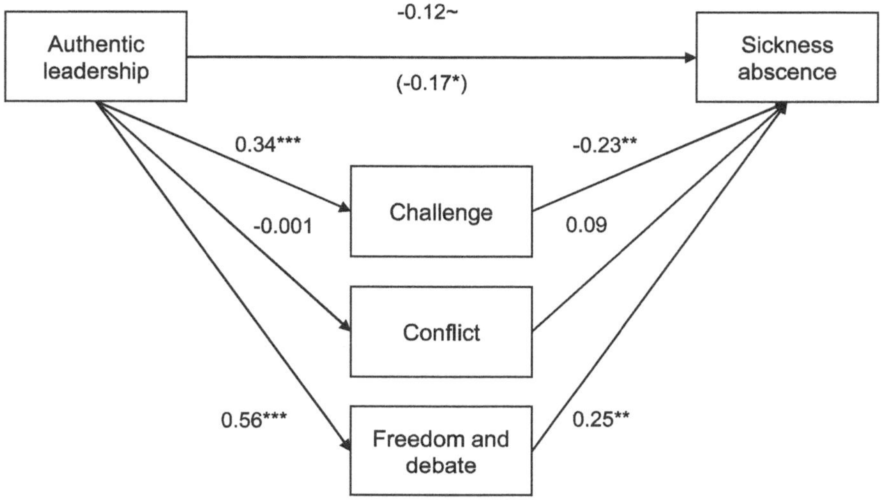

FIGURE 3.5 Standardized regression coefficients for the relationship between authentic leadership and worker sickness absence as mediated by climate for creativity (challenge, conflict, freedom and debate). ***$p < 0.001$. **$p < 0.01$. *$p < 0.05$.

TABLE 3.4
Estimation of Indirect Climate for Creativity Effects in the Relationship between Authentic Leadership and Sickness Absence among Workers ($N = 270$)

Indirect Effects B (BootSE)		95% Boot CI	
		BootLLCI	**BootULCI**
Total	0.06 (0.04)	−0.03	0.13
Authentic leadership -> challenge -> sickness absence	−0.08 (0.04)	−0.16	−0.001
Authentic leadership -> conflict -> sickness absence	0.00 (0.01)	−0.28	0.02
Authentic leadership -> freedom and debate -> sickness absence	0.20 (−0.09)	−0.01	0.36

Note: B = standardized indirect effect; BootSE = bootstrapped standard error; Boot CI = bootstrapped confidence interval; BootLLCI = bootstrapped lower limit of the confidence interval; BootULCI = bootstrapped upper limit of the confidence interval; level of confidence for the confidence intervals = 95%.

$^{***}p < 0.001$. $^{**}p < 0.01$. $^{*}p < 0.05$.

3.4 DISCUSSION OF RESULTS

In summary, both authentic leadership and the climate for creativity have proven to be important predictors of worker innovativeness and wellbeing. The results are consistent with the authentic leadership concept (Avolio et al. 2004) and the definition of the climate for creativity (Ekvall 1996).

There has been identified a strong, positive relationship between authentic leadership and worker innovativeness: the more authentic the manager is, the more innovative the employees turn out to be. Hence, the result proves that the authentic leadership yields greater work outcomes, not only in terms of worker performance (Avolio et al. 2004). The authentic leadership style has also been found indirectly related to worker innovativeness by two climate for creativity aspects: the *challenge* and the *freedom and debate*. However, this only partially supports the formulated hypothesis: the relationship between *challenge* and worker innovativeness has revealed a positive association, whereas the relationship between *freedom and debate* and worker innovativeness has proven negative. The positive relationship, consistent with the hypothesis, shows that managers influence worker innovativeness by creating appropriate psychosocial working conditions, i.e. the climate for creativity. This result combines two concepts: the relationship between job resources, organizational climate, and innovativeness (Ekvall 1996), and the concept of authentic leadership, revealing the way managers determine the performance of employees (Avolio et al. 2004). Managers who use the authentic leadership style create an atmosphere of a challenge faced by employees. An authentic leader aims at encouraging workers to develop their strengths, which can be expressed as a job challenge, such as a complex problem solution task, whereby employees are motivated to resolve the issue by being inspired by the authentic leader as a role model. Inciting workers to produce novel ideas and framing innovation as a job demand significantly increase innovative work outcomes. Such a working environment seems to remove psychosocial barriers to innovation in terms of both generating and implementing novel ideas (Hunter et al. 2007).

The negative relationship between *freedom and debate* and worker innovativeness has not supported the study hypothesis. This means that the *freedom and debate* was related to a reduced level of innovativeness. This result can be explained in several ways. Firstly, the *freedom and debate* aspect comprises three subscales: freedom, trust, and calm. Perhaps these subscales have a diverse impact on worker innovativeness, hence the negative result of the entire scale. Moreover, the *calm* subscale, related to Ekvall's "Idea time" factor, although related to creativity and innovate work outcomes (Karwowski 2009a; Ekvall 1996), may not have a positive impact on worker innovativeness. Innovation is more than creativity as it entails the entire innovation process, not only at the initial stage of creating a novel idea when the time for reflection is indeed required. At further stages of the innovation cycle, a certain energy may be essential, as mentioned by Karwowski (2009a). The organizational calm may also be perceived as a kind of stagnation that does not encourage creativity or innovation. Finally, according to Ekvall's concept (1996), a component of the climate for creativity is dynamism, corresponding to a lively, active organizational culture.

In turn, the positive relationships between the authentic leadership style and *freedom and debate,* and between the authentic leadership style and *challenge* seem straightforward – a manager who builds open, trust-based relationships with employees and appreciates employee potential and talent, also creates more opportunities for initiative and autonomy in performing work tasks. Authentic leaders thus encourage workers to develop skills, achieve personal fulfillment, and reach for more ambitious tasks, fostering a challenging workplace atmosphere.

Conversely, the authentic leadership style was not related to *conflict* – the third dimension of the climate for creativity. This means that authentic leaders have no influence over this particular factor in the workplace. Different interpretations are possible. Firstly, the author of the *climate for creativity* measurement tool highlights the lowest reliability of the *conflict* scale among the three *climate for creativity* scales and announces further work in this respect (Karwowski 2009a). Therefore, it is conceivable that the poor psychometric properties of the scale disturbed the results. Secondly, authentic leaders may not have such a strong influence on employees as to make them more self-aware or transparent, regardless of the circumstances. However, the evidence supports the stance that authentic, ethical leaders are more effective in solving workplace conflicts (Fotohabadi and Kelly 2018) and in helping workers to develop conflict management skills (Babalola et al. 2018).

Conflict was also significantly related to the *innovativeness*. However, it did not play a mediating role in the relationship between authentic leadership style and worker innovativeness. Contrary to the research hypothesis, this relationship was positive, i.e. the higher the *conflict* observed in the working environment, the more innovative the workers were. This result has two possible explanations. Firstly, the already mentioned poor psychometric properties of the scale may have disturbed the results. Secondly, the atmosphere of insecurity at work may have been conducive to the emergence of unconventional ideas among employees. This explanation has been supported by research proving that innovation is a tool to compete in a maximized performance race in insecure working conditions (Oke et al. 2012).

The survey results have proven that encouraging workers to realize their potential and develop their strengths, as well as creating open, trust-based relationships, a characteristic of the authentic leadership style, have a positive impact on the employee *work engagement*, a reduced *intention to quit the job* and *sickness absence*. This supports the accounts of the positive impact of the authentic leadership style on the health and wellbeing of workers, as well as work engagement and job satisfaction levels (Gardner et al. 2005; Bamford et al. 2013; Giallonardo et al. 2010; Dan-Shang et al. 2013). However, none of the climate for creativity aspects has acted as a mediator between authentic leadership and work engagement, contrary to the study hypothesis.

The relationship between the authentic leadership style and work-related stress has not been proven either. Similar results can be found in the literature, e.g. Rahimnia and Sharifirad (2015) demonstrated that the authentic leadership style was not directly related to stress levels. In the professional services sector, where there is high performance pressure, it is likely that workers would experience work- and performance-related stress, regardless of the first-line manager's behavior. Interestingly, an increased stress level was also related to the *challenge* – one of the climate for creativity aspects, but also a characteristic of the services sector where there is a strong emphasis on productivity. The atmosphere of a challenge faced by workers, greater tolerance for risk-taking behaviors, and complex problems to solve, all stimulate worker innovativeness on the one hand; however, they can be a potential source of stress on the other.

In turn, the more conflicts there were in the working environment, the greater the intent to quit the job reported by the workers, the more they experienced work-related stress, and the less engaged at work they were. Although there was observed no direct relationship between authentic leadership and work-related stress, the *challenge* and the *freedom and debate* played a mediating role: the more authentic the manager was, the more the employees perceived the organization as a challenging and open workplace, leading to higher (in relation to the *challenge*) and lower (in relation to the *freedom and debate*) work-related stress.

Sickness absence was related to two climate for creativity aspects: the *challenge* and the *freedom and debate*. The more the working environment was characterized by an atmosphere of a challenge, the less the workers were on sick leave. On the other hand, the higher the level of *freedom and debate* practiced at the workplace, the more frequent the sickness absence was observed. This result, although not supporting the study hypothesis, may have not corresponded to a poorer wellbeing of workers employed in companies where an atmosphere of freedom and debate was present. This equally may have proven that these workers were not afraid to take sick leave as there was no presenteeism culture in those workplaces (e.g. Ferreira et al. 2015). However, this is only a hypothesis, which should be treated carefully, as the respondents were not interviewed on performing work tasks during a period of illness, and thus the premise could not be verified. Nevertheless, it may be supported by the result showing that this factor was strongly related to a lower level of work-related stress among the workers. These factors also acted in some cases as partial mediators in the relationship between authentic leadership and sickness absence: the more authentic the manager was, the higher the *challenge* and the *freedom and debate* reported by the employees, resulting in lower and higher sickness absence, respectively.

When discussing the results of the conducted analyses, it should be mentioned that the tested models were confirmed to a varying degree. The worker innovativeness model was stronger, explaining 24% of the innovativeness level variance. The worker wellbeing models were characterized by a lower level of explained variance. While the remaining tested models explained 13% of the work engagement variance, and 12% of the work-related stress variance, the intention to quit the job and sickness absence were only explained within a 6–7% range. This suggests that more tangible indicators of worker wellbeing are explained by other variables, or over a longer term, which could be observed in longitudinal studies.

3.5 STRENGTHS AND LIMITATIONS

The strength of the present study has been the combination of two models that capture the impact of the work environment on the employee: the authentic leadership model and the climate for creativity model. The results have shown that the same variables that are responsible for worker wellbeing are also related to worker innovativeness, suggesting that a healthy worker is an essential part of a healthy organization and that a healthy organization supports worker health. In view of the inconsistent results produced by international research so far, it is also important that the study was conducted in Poland, where in recent years attention has been increasingly paid to positive occupational psychology, and the determinants of employee work engagement and innovativeness.

However, there were some limitations to the current study. The obtained results of the conducted analyses should be interpreted with caution, as the survey was carried out as a cross-sectional study. Accordingly, the mediation analyses' results should be particularly treated with caution. Another limitation was the measurement of variables based on a single source, i.e. the surveyed workers. Therefore, a common method bias might have taken place, which should also be taken into account when interpreting the results.

3.6 CONCLUSIONS

In conclusion, the authentic leadership style seems to be generally related to a greater innovativeness and wellbeing of workers, and these relationships are partly explained by the climate for creativity. However, not all the results have been consistent with the study hypotheses here discussed. Nevertheless, the study results suggest that a work environment which provides workers with engaging, meaningful work as well as thought-provoking, and dialogue promoting climate is essential if workers are to be inspired to produce innovation and enjoy good health and wellbeing.

REFERENCES

Aarons, G. A., and D. H. Sommerfeld. 2012. Leadership, innovation climate, and attitudes toward evidence-based practice during a statewide implementation. *Journal of the American Academy of Child & Adolescent Psychiatry* 51(4):423–431.

Amabile, T. M., and N. D. Gryskiewicz. 1989. The creative environment scales: Work environment inventory. *Creativity Research Journal* 2(4):231–253.

Avolio, B. J., W. L. Gardner, F. O. Walumbwa, F. Luthans, and D. May. 2004. Unlocking the mask: A look at the process by which authentic leaders impact follower attitudes and behaviors. *Leadership Quarterly* 15(6):801–823. DOI: 10.1016/j.leaqua.2004.09.003.

Avolio, B. J., W. L. Gardner, and F. O. Walumbwa 2007. Authentic leadership questionnaire (ALQ). http://www.mindgarden.com. (accessed 28 January 2020).

Babalola, M. T., J. Stouten, M. C. Euwema, and F. Ovadje. 2018. The relation between ethical leadership and workplace conflicts: The mediating role of employee resolution efficacy. *Journal of Management* 44(5):2037–2063.

Bamford, M., C. A. Wong, and H. Laschinger. 2013. The influence of authentic leadership and areas of worklife on work engagement of registered nurses. *Journal of Nursing Management* 21(3):529–540.

Barrett, H., J. Balloun, and A. Weinstein. 2005. The impact of creativity on performance in nonprofits. *International Journal of Nonprofit & Voluntary Sector Marketing* 10(3):213–223.

Burpitt, W. J., and W. J. Bigoness. 1997. Leadership and innovation among teams: The impact of empowerment. *Small Group Research* 28(3):414–423.

Černe, M., M. Jaklič, and M. Škerlavaj. 2013. Authentic leadership, creativity, and innovation: A multilevel perspective. *Leadership* 9(1):63–85.

Dan-Shang, W., and H. Chia-Chun. 2013. The effect of authentic leadership on employee trust and employee engagement. *Social Behavior & Personality: An International Journal* 41(4):613–624.

Denti, L., and S. Hemlin. 2012. Leadership and innovation in organizations: A systematic review of factors that mediate or moderate the relationship. *International Journal of Innovation Management* 16(03):1–20. DOI: 10.1142/S1363919612400075.

De Jong, J., and D. Den Hartog. 2010. Measuring innovative work behaviour. *Creativity and Innovation Management* 19(1):23–36.

Donaldson-Feilder, E., F. Munir, and R. Lewis 2013. Leadership and employee well-being. In: *The Wiley-Blackwell Handbook of the Psychology of Leadership, Change, and Organizational Development*, eds. H. S. Leonard, R. Lewis, A. M. Freedman, and J. Passmore, 155–173. Wiley-Blackwell. DOI: 10.1002/9781118326404.ch8.

Ekvall, G. 1996. Organizational climate for creativity and innovation. *European Journal of Work & Organizational Psychology* 5(1):105–123.

Ekvall, G., and L. Ryhammar. 1999. The creative climate: Its determinants and effects at a Swedish university. *Creativity Research Journal* 12(4):303–310.

Elo, A. L., A. Leppänen, and A. Jahkola. 2003. Validity of a single-item measure of stress symptoms. *Scandinavian Journal of Work, Environment & Health* 29(6):444–451.

Elrehail, H., O. L. Emeagwali, A. Alsaad, and A. Alzghoul. 2018. The impact of transformational and authentic leadership on innovation in higher education: The contingent role of knowledge sharing. *Telematics & Informatics* 35(1):55–67.

Ferreira, A., L. Martinez, C. Cooper, and D. Gui. 2015. LMX as a negative predictor of presenteeism climate: A cross-cultural study in the financial and health sectors. *Journal of Organizational Effectiveness: People & Performance* 2(3):282–302.

Fotohabadi, M., and L. Kelly. 2018. Making conflict work: Authentic leadership and reactive and reflective management styles. *Journal of General Management* 43(2):70–78.

Gardner, W. L., B. J. Avolio, F. Luthans, D. R. May, and F. O. Walumbwa. 2005. Can you see the real me? A self-based model of authentic leader and follower development. *Leadership Quarterly* 16(3):434–372.

Giallonardo, L. M., C. A. Wong, and C. L. Iwasiw. 2010. Authentic leadership of preceptors: Predictor of new graduate nurses' work engagement and job satisfaction. *Journal of Nursing Management* 18(8):993–1003.

Gumusluoğlu, L., and A. Ilsev. 2009. Transformational leadership and organizational innovation: The roles of internal and external support for innovation. *Journal of Product Innovation Management* 26(3):264–277.

Hayes, A. F. 2009. Beyond Baron and Kenny: Statistical mediation analysis in the new millennium. *Communication Monographs* 76(4):408–420.

Hayes, A. F. 2017. *Introduction to Mediation, Moderation, and Conditional Process Analysis. Gression-Based Approach*. 2nd Edition. New York: Guilford Press.

Herrmann, D., and J. Felfe. 2014. Effects of leadership style, creativity technique and personal initiative on employee creativity. *British Journal of Management* 25(2):209–227.

Hunter, S. T., K. E. Bedell, and M. D. Mumford. 2007. Climate for creativity: A quantitative review. *Creativity Research Journal* 19(1):69–90.

Ismail, M. 2005. Creative climate and learning organization factors: Their contribution towards innovation. *Leadership & Organization Development Journal* 26(8):639–654.

Jung, D. I. 2001. Transformational and transactional leadership and their effects on creativity in groups. *Creativity Research Journal* 13(2):185–195.

Karwowski, M. 2009a. *Klimat dla kreatywności. Koncepcje, metody, badania*. Warszawa: Wydawnictwo Difin.

Karwowski, M., and K. Pawłowska. 2009b. Klimat dla kreatywności w miejscu pracy. *Bezpieczeństwo Pracy – Nauka i Praktyka* 2(449):18–22.

Kauppinen, T., R. Hanhela, P. Heikkila et al. 2007. *Tyo ja terveys Suomessa 2006*. [Work and Health in Finland 2006]. Helsinki: Finnish Institute of Occupational Health.

Kim, M. 2018. The effects of authentic leadership on employees' well-being and the role of relational cohesion. In: *Leadership*, ed. Suleyman D. Göker, IntechOpen. DOI: 10.5772/intechopen.76427.

Laschinger, H. K. S., and R. Fida. 2014. New nurses burnout and workplace wellbeing: The influence of authentic leadership and psychological capital. *Burnout Research* 1(1):19–28.

Matzler, K., E. Schwarz, N. Deutinger, and R. Harms. 2008. The relationship between transformational leadership, product innovation and performance in SMEs. *Journal of Small Business & Entrepreneurship* 21(2):139–151.

McMurray, A. J., A. Pirola-Merlo, J. C. Sarros, and M. M. Islam. 2010. Leadership, climate, psychological capital, commitment, and wellbeing in a non-profit organization. *Leadership & Organization Development Journal* 31(5):436–457. DOI: 10.1108/01437731011056452.

Munir, F., K. Nielsen, A. H. Garde, K. Albertsen, and L. G. Carneiro. 2012. Mediating the effects of work–life conflict between transformational leadership and health-care workers' job satisfaction and psychological wellbeing. *Journal of Nursing Management* 20(4):512–521.

Nęcka, E. 2001. *Psychologia twórczości*. Gdańsk: Gdańskie Wydawnictwo Psychologiczne.

Nelson, K., J. S. Boudrias, L. Brunet et al. 2014. Authentic leadership and psychological wellbeing at work of nurses: The mediating role of work climate at the individual level of analysis. *Burnout Research* 1(2):90–101.

Nyberg, A., P. Bernin, and T. Theorell 2005. The impact of leadership on the health of subordinates Report no 1. Stockholm: SALTSA. http://www.su.se/polopoly_fs/1.51750.1321891474!/P2456_AN.pdf (accessed 28 January 2020).

Nyberg, A., H. Westerlund, L. L. Magnusson Hanson, and T. Theorell. 2008. Managerial leadership is associated with self-reported sickness absence and sickness presenteeism among Swedish men and women. *Scandinavian Journal of Public Health* 36(8):803–811.

Oke, A., F. O. Walumbwa, and A. Myers. 2012. Innovation strategy, human resource policy, and firm's revenue growth: The roles of environmental uncertainty and innovation performance. *Decision Sciences* 43(2):273–302.

Paulsen, N., D. Maldonado, V. J. Callan, and O. Ayoko. 2009. Charismatic leadership, change and innovation in an R&D organization. *Journal of Organizational Change Management* 22(5):511–523.

Purkiss, R. B., and R. J. Rossi 2008. Sense of community: A vital link between leadership and wellbeing in the workplace. In: *Advances in Organisational Psychology*, eds. A. Glendon, B. M. Thompson, and B. Myors, 281–296. Bowen Hills: Australian Academic Press.

Rahimnia, F., and M. S. Sharifirad. 2015. Authentic leadership and employee wellbeing: The mediating role of attachment insecurity. *Journal of Business Ethics* 132(2):363–377.

Schaufeli, W. B., and A. B. Bakker 2004. UWES – Utrecht work engagement scale: Preliminary manual. Utrecht: Occupational Health Psychology Unit, Utrecht University. https://www.wilmarschaufeli.nl/publications/Schaufeli/Test%20Manuals/Test_manual_UWES_English.pdf (accessed 29 January, 2020).

Schaufeli, W. B., A. Shimazu, J. Hakanen, M. Salanova, and H. De Witte. 2017. An ultra-short measure for work engagement: The UWES-3 validation across five countries. *European Journal of Psychological Assessment* 35(4):577–591.

Sellgren, S. F., G. Ekvall, and G. Tomson. 2008. Leadership behaviour of nurse managers in relation to job satisfaction and work climate. *Journal of Nursing Management* 16(5):578–587.

Shrout, P. E., and N. Bolger. 2002. Mediation in experimental and nonexperimental studies: New procedures and recommendations. *Psychological Methods* 7(4):422.

Skakon, J., K. Nielsen, V. Borg, and J. Guzman. 2010. Are leaders' well-being, behaviours and style associated with the affective well-being of their employees? A systematic review of three decades of research. *Work & Stress* 24(2):107–139.

Tepper, B. J. 2000. Consequences of abusive supervision. *Academy of Management Journal* 43(2):178–190.

West, M. A. 2002. Sparkling fountains or stagnant ponds: An integrative model of creativity and innovation implementation in work groups. *Applied Psychology: An International Review* 51(3):355–387.

West, M. A., C. S. Borrill, J. F. Dawson, F. Brodbeck, D. A. Shapiro, and B. Haward. 2003. Leadership clarity and team innovation in health care. *The Leadership Quarterly* 14(4–5):393–410.

Widerszal-Bazyl, M. 2003. *Stres w pracy a zdrowie – Czyli o próbach weryfikacji modelu Roberta Karaska oraz modelu: wymagania-kontrola-wsparcie.* Warszawa: CIOP-PIB.

Wiezer, N., K. Nielsen, K. Pahkin et al. 2011. *Exploring the Link between Restructuring and Employee Wellbeing.* Warszawa: CIOP-PIB.

Zhou, J., Y. Ma, W. Cheng, and B. Xia. 2014. Mediating role of employee emotions in the relationship between authentic leadership and employee innovation. *Social Behavior & Personality: An International Journal* 42(8):1267–1278.

4 Selected Employment Characteristics and Employee Health and Performance

The Mediating Role of the Psychological Contract

Dorota Żołnierczyk-Zreda

CONTENTS

4.1 INTRODUCTION

Over the past few decades the nature of work has evolved toward more non-standard forms of employment, known as a contingent, alternative, atypical, or non-permanent work (Connelly and Gallagher 2006; Kalleberg 2009; Kalleberg 2011; ILO 2016).

Various countries have different temporary employment regulations. For example, European law protects temporary workers to a greater extent than the Australian, Canadian, and American regulations (Zeytinoglu and Cooke 2005; Guest 2004; Kallenberg 2009; OECD 2019).

According to the ILO classification, non-standard employment encompasses the following four forms of employment:

1. Temporary employment (e.g. fixed-term employment, project-based employment, seasonal work)
2. Part-time employment (half-time/part-time work, zero hours contract, on-call work)
3. Multiple parties and contractual relationships employment (leased, networked, sub-contracted, temporary agency work)
4. Disguised employment relationships and dependent self-employment (ILO 2016)

The number of employees working in non-standard forms of employment has been growing as a result of globalization processes, increased competition, and dynamic technological development, including digitalization, which all push for enterprise innovation and continuous restructuring, thus also increasing the need to manage labor costs and enterprise flexibility (Kalleberg 2009; Katz and Krueger 2016). This flexibility is typically defined as enterprise ability to adapt to customer requirements (Wood 2016).

At the turn of the 21st century the percentage of persons working in non-standard forms of employment grew faster in Europe (14%), G7 countries (6.7%), and OECD countries (10.7%) than in the U.S. (4.7%) (Cappelli and Keller 2013, OECD, 2019). However, in 2005 it already amounted to 10.7%, and in 2015 it rose to 15.8% in the U.S. (Katz and Krueger 2016). American researchers identify two types of non-standard/alternative forms of employment: one refers to high-skilled workers who engage in atypical employment by volition; and the other relates to low-skilled workers who struggle to make ends meet and have to adjust to employer needs. The first category also includes persons who make up a small percentage of all employees (0.5%), but are the fastest-growing type of alternative employment workers, performing technology platform-based and intermediary work, such as Uber, Airbnb, or Lyft online services, so-called *gig work* (Katz and Krueger 2016; Farrell and Grig 2016).

In Europe, the largest non-standard employment segment represents various types of temporary work (Eurofound 2018). It has grown significantly since the mid-1980s, in France and Spain between 1985 and 1995, in Sweden in the early 1990s, and in Germany in the early 2000s.

There has been no upward trend in the total number of temporary employment contracts in the European Union in the last decade, however, the decline in this area has been insignificant (from 14.5% in 2006 to 14.1% in 2018). Young workers of foreign origin, persons with a low education level, and those in basic-skills jobs are more likely to have a fixed-term contract. At present, the gender gap is less than one percentage point. Averaged statistical data on temporary workers show that their socio-economic status is far worse off, although the outlook for Europe reveals some complexity in this area. Non-standard employment workers have increasingly been entering *the precariat* – a growing pool of workers experiencing job insecurity, working in conditions of reduced social security entitlements, low wages, and no professional competence-raising opportunities, all of which make for a volatile and precarious livelihood (Standing 2011; Benach et al. 2014). According to the Eurofound *Precarious Employment in Europe 2015* report (Eurofound 2015), the highest precarious employment rates were noted for Poland, Bulgaria, Estonia, Greece, Latvia, Lithuania, and Spain. Furthermore, in Central and Eastern European countries, such as Poland or Lithuania, there are no robust social dialogue structures or collective bargaining practices, a fact which exacerbates the insecurity of temporary employment. In addition, the rate of transition from temporary to permanent employment declined from 28% in 2005–2006 to 20% in 2011–2012, with rates lower than 20% recorded for France, the Netherlands, Spain, Greece, Italy, and Poland, suggesting a labor market segmentation in these countries (Eurofound 2015).

4.1.1 Fixed-Term Employment in Poland

A large growth of temporary employment in Poland took place in the first decade of the 21st century, from only 4.6% in 1999 to 28.2% in 2007. This rate was sustained up to 2016, recording the highest rise of fixed-term employment in Europe between 2010 and 2016. In 2018, Poland 'dropped' to third place in the European fixed-term employment statistics, just behind Montenegro and Spain, with 24.3% of the entire workforce employed on fixed-term contracts, compared to the European average of 14.1% and OECD average of 11% (OECD Employment Outlook, 2019).

The decline occurred in the aftermath of *The Great Social Debate* – a public discussion on the negative consequences of fixed-term employment, followed by amendments to the Labor Code in 2016. According to the amended regulations, an employer may be kept on up to three successive fixed-term contracts with the same employer for a limit of three years. The total length of the fixed-term employment shall not exceed 33 months. Temporary employment has been considered a less precarious form of work ever since, however, it can still be terminated by the employer without any formal notice, or justification. Existing data show that temporary workers are mostly employed in the following three industry sectors: professional services, construction, and manufacturing, making up mainly customer service and sales workers, operators and machinery assemblers, and workers in low-skill jobs, respectively (Cichocki et al. 2013; Chłoń-Domińczak and Pałczyńska 2015).

In Poland, there have been observed low temporary to permanent contract transition rates, since only one-third of temporary workers obtain a permanent contract after the three-year period of legal temporary employment has passed. The likelihood of transitioning to permanent employment is mainly tied to age and education level, the highest chances being noted for prime-age, university-educated employees, and the lowest for persons with a primary school or lower secondary education, regardless of age.

As temporary-contract employees rarely perform high-skill tasks, they are less likely to take part in employer-provided training schemes. Hence, temporary workers not only have fewer opportunities to apply their skills at work, but also have lower chances of developing professional competencies, and thus accumulating individual human capital, considered a key prerequisite in securing a permanent job contract (Chłoń-Domińczak and Pałczyńska 2015; Pańków 2015). Moreover, temporary-contract employees report lower earnings, and such a discrepancy has persisted over time (Lewandowski et al. 2017).

4.1.2 Type of Employment and Employee Well-Being

The growing scale of fixed-term employment in Europe and worldwide, in particular the precariousness attributed to temporary work, has prompted many researchers to examine whether this type of employment has been associated with poorer physical and mental health of employees.

Although an initial meta-analysis of data on the relationship between temporary employment and employee well-being conducted by Virtanen et al. in 2005 revealed greater psychological problems among temporary workers than permanent workers, these differences were not significant. However, it is worth mentioning that different risk factors' definitions, mental health indicators, and contextual factors have been considered by researchers in the field. Virtanen et al. (2001) obtained inconsistent results in a study of medical staff at ten hospital establishments in Finland. The aim of the study was to compare the health status of temporary and permanent employees. The results revealed that temporary agency workers reported a better self-assessment of general health status than permanent employees. Moreover, there were no differences in the prevalence of diagnosed chronic diseases and psychiatric problems between the two study groups. In another study, Pekka Virtanen and colleagues (2003b) examined health differences among 15,468 permanent and atypical employment workers, including temporary workers, unemployed persons, and job seekers. Fixed-term workers were found to have a similar number or even fewer health problems (related to general health and depression) than permanent workers (Virtanen et al. 2003). Mixed conclusions on the health of temporary agency workers were also shown in several other studies (Aronsson et al. 2002; Bardasi and Francesconi 2004; Artazcoz et al. 2005).

Subsequent studies, particularly longitudinal ones, generally supported the association between temporary employment and poor mental health in European and non-European populations. A study carried out in Finland on a large sample of 107,828 Finnish public sector employees revealed that depression-related episodes of work incapacity were longer among temporary workers, compared to permanent workers. This factor prolonged the return to work, particularly among ageing adults and low-educated persons (Ervasti et al. 2014). In turn, Vives et al. (2013) surveyed 5,679 temporary and permanent Spanish workers and observed that those with short-term employment contracts were more likely to be depressed than persons with long-term employment contracts. These researchers also concluded that the lower the level of employment stability, the worse the mental health of employees.

The impact of precarious employment on mental health was also analyzed on a very large sample of over 2.7 million employees in Italy (Moscone et al. 2016). The results revealed that the probability of taking psychotropic drugs was higher for temporary workers. Longer working hours under temporary contracts increased the likelihood of mental health problems that required pharmaceutical treatment. The researchers also noted that the transition from permanent to temporary employment also increased the incidence of such mental health problems.

The relationship between temporary employment and poor employee well-being has also been supported by several South Korean studies. For instance, Kim et al. (2008) proved that precarious employment workers, defined in the study as workers employed on a temporary, part-time, or daily basis, or in a contingent (fixed short-term) employment, were in poorer health than permanent workers. A subsequent study by Han et al. (2017) of 24,173 Korean workers proved that precarious employment was significantly associated with depressive moods, including suicidal ideation, but only among male workers and workers with low to medium income levels. The precarious work category encompassed in the study a wide range of employees, from temporary, daily, part-time, dispatched, and subcontracted workers to other atypical employment workers.

Quasnel-Vallée et al. (2010) analyzed the *U.S. National Longitudinal Survey of Youth* data and observed that temporary employment at a younger age increased the risk of depressive disorders later on in life, however, only for temporary employment agency workers and sub-contractors.

Several other sources of data have shown that temporary agency work is more harmful to the mental health of women (Kim et al. 2016), or men (Han et al. 2017; Kachi et al. 2014), or only to employees with a lower education level (Arctotoz et al. 2005, Hammarström 2011).

Nevertheless, there are also results suggesting that ill health, including mental health conditions, apply to permanent rather than temporary employees. For example, Guest, Isaksson and De Witte (2010) surveyed over 5,000 permanent and temporary workers from Sweden, Germany, the Netherlands, Belgium, the UK, Spain, and Israel. Most of the temporary workers were on fixed-term contracts with an employment agency, the remaining were seasonal workers. Contrary to the study hypothesis, the researchers revealed that permanent workers, rather than temporary staff, experienced poorer well-being, i.e. higher levels of irritation, anxiety, and depression, and lower levels of job satisfaction, coupled with a poorer general health status. Similarly, Goudswaard and Andries (2002) analyzed indicators of the well-being and working conditions of temporary agency workers in Denmark, Finland, France, Germany, the Netherlands, Spain, and Sweden. The researchers revealed that temporary agency workers had lower levels of stress and fewer musculoskeletal disorders than permanent workers.

4.1.3 Type of Employment, Work Engagement, and Employee Performance

In a number of studies researchers have also attempted to investigate whether temporary agency workers differ from permanent workers in terms of work engagement and performance. The results, similar to the research on the relationship between the type of employment contract and well-being, have also been mixed. Some results have proved that persons employed on fixed-term contracts are less engaged at work than persons employed on permanent contracts (Coyle-Shapiro and Kessler, 2002; De Gilder 2003; Guest 2004; De Jong and Schalk 2005; Rigotti and Mohr 2005). Other findings have not confirmed the negative relationship between temporary employment and work engagement (De Witte and Näswall 2003; Van Breukelen and Allegro 2000), and have even revealed lower work engagement among permanent employees.

For example, Allen concluded that Australian fixed-term contract nurses had lower levels of engagement and a less optimistic outlook on future career prospects than permanent-contract nurses (Allen 2011). Similarly, among Portuguese workers with various employment contracts, only workers with permanent and fixed-term contracts were strongly engaged with the organization, whereas temporary agency workers did not display such behavior (Chambel and Castanheira 2006). This highlights the dual (and different) commitment of temporary agency workers: toward the agency and toward the employer (Van Breugel et al. 2005). While the employee commitment to the employer is a stable relationship, the commitment to the agency depends on the number of job offers it secures, and the duration of the contract with the agency (Gallagher and Parks 2001).

In studies conducted by Coyle-Shapiro and Kessler (2002), De Gilder (2003), and Guest (2004) the results proved that permanent-contract employees manifested a higher level of organizational citizenship behavior than temporary-contract employees. Conversely, Guest, Isaksson, and de Witte (2010) noted that permanent workers did not differ from temporary workers in terms of work engagement. This was also acknowledged by HR managers who declared they were equally satisfied with the performance of both employee groups.

In relation to the latter indicator of work attitudes, namely the performance of temporary agency workers, research findings are also inconsistent. Engellandt and Riphahn (2005) surveyed Swiss workers and revealed that temporary-contract workers made a greater effort at work than permanent workers. The authors also revealed that the former were ready to devote about 60% more time to unpaid overtime than permanent-contract employees, although, this was true only for those temporary employees who had promotion prospects. The authors did not observe, however, the effect of lower sickness absence among temporary workers, which is often (in Switzerland) considered a way of signaling dissatisfaction with the workplace.

Meyer and Wallette (2005), analyzed a sample of 360,000 employees included in the Swedish *Labour Force Surveys* (LFS) and revealed that although Swedish temporary workers had lower absenteeism rates than permanent employees, they were less likely to work after hours. Sverke et al. (2002) observed that the job insecurity experienced by temporary employees led to a decrease in work engagement, lack of trust in the organization, resistance to organizational change, and reduced performance. Kalleberg (2001) suggests that temporary workers are characterized by lower performance because they are newcomers and are yet to learn both the work and the organization. De Cuyper and De Witte (2006a) and Ellingson et al. (1998) did not confirm significant differences in the performance of permanent and temporary workers, while Van Breukelen and Allegro (2000) observed that managers were more satisfied with the performance of temporary workers than permanent employees.

4.1.4 Employment Contract Volition and Employee Health and Performance

The temporary employment contract of choice is considered an important predictor of the health and performance of temporary workers (Connelly and Gallagher 2006). The International Labour Organization also recognizes the principle of voluntary temporary work as the second of the three most important criteria of (high) quality of work (ILO 2016). According to the ILO definition, involuntary temporary-contract employment applies to employees who are unable to find equivalent permanent-contract work (ILO 2016).

Previous research results have confirmed that temporary agency workers with voluntary contracts enjoy greater job satisfaction (Ellingson et al. 1998; Krausz et al. 2000), are more engaged at work, and are healthier than temporary agency workers who would prefer a different (generally permanent) type of contract (Isaksson and Bellaagh 2002).

The holding of temporary employment may be motivated by various reasons. It is believed, for example, that temporary work provides a better opportunity to maintain a healthy work–life balance (Ellingson et al. 1998; Tan and Tan 2002). It may also offer temporary workers greater autonomy in pursuing their career path by reducing employer control over employees. Equally, it may increase employee-perceived freedom as changing jobs is easier on temporary contracts (Lopes and Chambel 2017). Temporary employment may also, particularly for young workers, constitute a chance of trying out different activities, and thus gaining work experience. Similarly, Aronsson and Göransson (Aronsson 1999) have proved that the choice of temporary employment may be related to the chosen career path in a preferred profession, or to a company that is renowned on the labor market and that such factors become more important than the type of employment.

Much attention has been paid in the existing literature on temporary workers to the concept of temporary employment as a *stepping stone* to permanent employment. It is believed that such motivation is tantamount to accepting temporary employment and is not associated with poorer health, well-being, and performance (van den Berg et al. 2009; Connelly and Gallagher 2006; De Cuyper 2006a; De Jong et al. 2009). Even if employees have an involuntary temporary employment contract they may have a strong motivation to work, hoping to obtain permanent posts.

DeCuyper and DeWitte (De Cuyper 2007b) revealed in their research that temporary workers who preferred temporary employment had a significantly higher level of job satisfaction and a lower level of irritation and intention to quit the job than those for whom temporary employment was an involuntary work contract.

At the same time, an interesting result in terms of preferences related to the type of contract was obtained in a study by Aronsson and Goransson (Aronsson 1999) conducted among 1,564 Swedish workers employed on various types of employment contracts. An unexpected survey result revealed that 28% of the surveyed workers employed on permanent contracts did not work in their preferred professions, and as many as 25% would have preferred temporary work if it were in their profession of choice. Among those who had temporary contracts and performed jobs which were not their professions of choice, only 52% would have liked to obtain permanent contracts. For half of the respondents, it was more important that the occupation/job, rather than the type of contract, was consistent with their choice.

It is believed that the motivation to move from temporary to permanent employment is based on the '*foot-in-the-door*' strategy. It is, similarly to the *stepping stone* motivation, associated with the acceptance of poorer working conditions by temporary workers and their high performance (De Cuyper et al. 2008c).

De Jong and Schalk (2010) showed, however, that an involuntary temporary employment contract proved to have a negative impact on the employee well-being and was more of a trap than a 'bridge' to better-quality employment. Ellingson et al. (1998) compared the performance of voluntary and involuntary temporary workers and found no significant differences between the groups.

Due to the higher rate of involuntary temporary employment recorded in Poland, compared to other countries, and the paucity of research on the relationship between the employment type preference and employee health and performance, the present research is an attempt to answer the question of the nature of this relationship.

4.1.5 Type of Work and Employee Health and Performance

The better physical health of employees performing non-manual work compared to those who perform manual work has been proved by much research (Vahtera et al. 1999; Lahelma et al. 2012). Hu et al. (2016) in a study conducted in 17 European countries showed that persons with lower levels of education and those performing manual work report poorer subjective health assessment.

Lahelma et al. (2012) concluded on the basis of their research that the physical strain experienced by low-skilled employees in their work was the main cause of their poorer health status, compared to the health status of high-skill workers. Numerous studies have also confirmed the lower work ability of the former employee group (Aittoma¨ki 2003; Lunde et al. 2014; Oliv et al. 2017). Thorsen et al. (Thorsen 2013) investigated the following predictive factors of sickness absence and low work ability in a representative sample of Danish workers ($N = 6.743$), performing both physical and intellectual work: age, social status, work ability, ergonomic risk exposure, noise, management support, and job control. They found that older age, poor health, low social status, and various physical risks at work were associated with a low work ability. The authors of the study concluded that the physical environment was a stronger predictor of work ability and sickness absence than the psychosocial environment. Leinonen et al. (Leinonen 2011) examined whether social status, health, and working conditions posed an early work incapacity risk among Finnish workers and found that the risk was significantly higher among workers with a lower social status, mediated by physical strain among female respondents, and exposure to hazardous factors at work among the surveyed men.

In a similar study by Lahelma et al. (Lahelma 2012), which also examined the working environment risk factors related to premature retirement as a result of work incapacity among Finnish workers, it was found that the most significant factor was the physical strain experienced by women and men, and the low job control noted among women. Similarly, Alavinia et al. (Alavinia 2008) revealed that in the group of more than 5,000 construction workers, poorer work ability was observed not only among those who carried loads, but also among those workers who had low job control.

In a study by Feldt et al. (Feldt 2009) the managerial position was not found to prevent a deteriorating work ability among these persons.

Section 4.1.2 presents research showing that temporary work is more harmful only to employees with a low level of education who perform manual work. Arctotoz et al. (Arctotoz 2005) found no direct relationship between temporary employment and poor health, and job satisfaction among the surveyed Spanish workers; however, the association was identified among men performing physical work. Hammarström et al. (Hammarström 2011) observed that among Finnish temporary workers, those with a low level of education and performing low-skill jobs had significantly poorer health status, comparing to those with a high level of education and performing more complex intellectual work tasks.

Marler et al. (Marler 2002) proved in her research that many persons (particularly young people) intentionally chose temporary employment as an opportunity to gain experience in multiple companies and learn new skills that could be easily transferred to other workplaces. This motivation is sometimes referred to as 'independent career orientation', but it usually concerns a number of occupations such as accountants, lawyers, engineers, and advertising professionals. For these people, working in many places is an enriching experience and enhances their professional career prospects. However, for persons with lower qualifications and social status performing manual work, temporary employment is often a 'dead end' on the career path. Those authors, in concluding their research results, were the first to postulate that further research efforts aimed at investigating the health and occupational consequences of temporary employment should be based on the distinction between two important categories of temporary workers. The first category would concern the occupational status of temporary workers, or the type of work they do. The second would be the volition in choosing this type of employment contract.

While the relationship between physical work and employee health has been a subject of much research, there are no data in the published literature on the differences between persons performing manual and non-manual work with regard to their work attitudes, i.e. work engagement and performance. There are no such data available in Poland either. Obtaining an answer to such a question has therefore been formulated as one of the present study aims.

4.1.6 The Mediating Role of Psychological Contract in the Relationship between Type of Employment Contract, Contract Volition, Type of Work Performed, and Employee Health and Performance

The aforementioned research results indicate that the type of employment (permanent or temporary), the employment contract preferences, as well as the nature of the work performed affect the health and performance of employees as a result of working conditions, specific to the type of employment. In the context of temporary employment, the term psychological contract is often used to describe working conditions (Guest 2004; De Cuyper et al. 2006a; De Cuyper et al. 2008b; De Jong et al. 2009; De Cuyper et al. 2011; Callea et al. 2016). With regards to psychological contract, the 'contract' is a metaphor derived from a legal employment contract, specifying the employer and employee obligations, i.e. a set of unwritten mutual expectations between the two parties (Schein 1978). This concept, developed intensively in organizational psychology, is used to describe mutual (also developing over time) formal and less formal aspects of employer–employee relations (Rousseau 1995; Rousseau et al. 2018; Conway and Briner 2005; Lambert 2011; Alcover et al. 2017). The more formal employer obligations that make up the psychological contract are of a financial nature (e.g. salary, fringe benefits) and constitute the transactional contract. Less formal employer commitments (e.g. support in personal matters, employee participation in decision making, career development opportunities, training, organizational policies and procedures enhancing task performance) relate to the psychosocial dimension and are referred to as relational contracts (Rousseau 2004). While transactional contracts focus on the short-term, financial liabilities of the employer and, at the same time, employee rights, relational contracts contain both financial and non-financial liabilities. Furthermore, the relational contract is based on trust, goodwill, and ongoing reciprocity in the exchange of these 'subtle goods' and the long-term loyalty of the employee to the employer (Rousseau 1990).

It is considered that the legal contract, and in particular the fixed-term employment contract, impacts on the psychological contract in two ways: firstly, it may limit the possibility of negotiating the psychological contract in order to extend its obligations. Secondly, the temporary employment contract strictly defines the duration of the mutual relationship between the employee and the employer, which also significantly affects the quality of the psychological contract (Rousseau 1995; McLean Parks et al. 1998). The fixed-term duration of a legal contract is therefore generally limited to focusing on the transactional elements of the psychological contract, whereas a permanent contract creates the possibility to build more relational aspects of the contract. These conclusions have been supported by empirical studies. Several studies have revealed poorer psychological contract obligations in relation to temporary workers, compared to their permanent counterparts. An example is the study conducted by De Jong and Schalk (2010), which proved that both employees and employers believed that the psychological contract obligations of temporary workers were fewer than those of permanent employees, and that the initial temporary contract commitments were more likely to be unfulfilled than the contractual agreements of permanent workers. The survey also showed that only the employer compliance with a majority of contract obligations was associated with a high job satisfaction, perception of fairness, and lower intention to quit the job among employees.

Much research on psychological contract has referred to the degree of psychological contract fulfillment vs. psychological contract breach as two extremes of a single continuum (Robinson and Rousseau 1994; Tekleab and Taylor 2003; Hekman et al. 2009; Conway and Coyle-Sharpiro 2012). Zhao et al. (Zhao 2007) have revealed in their meta-analysis of previous research on the psychological contract breach that it is associated with a lack of trust in the organization, lower job satisfaction, less work engagement, greater intention to quit the job, and falling work performance.

It is believed that employer failure to fulfill psychological contracts of temporary workers may prove that they have little chance of having their contracts renewed, resulting in increased job insecurity and poorer well-being of such employees (De Witte and Näswall 2003; De Cuyper 2006a; De Cuyper and Witte 2007a). According to these authors, permanent workers are less susceptible to psychological contract breach.

The psychological contract has been considered a very significant, often even more important than the legal contract, predictor of mutual relations between employer and employee, job satisfaction, and work engagement (De Cuyper, et al. 2011). The approach was pioneered by British researchers (Guest and Clinton 2006), who examined whether the psychological contract was a mediator of the impact of the employment contract type on numerous health and organizational behavior indicators among permanent and temporary employees. Health was assessed as symptoms of work-related anxiety and depression, perceived self-efficacy, job satisfaction, and work–life balance, whereas the following were organizational behavior indicators: work performance, absenteeism, accidents, intention to quit the job. In the theoretical model tested by the authors, apart from the type of employment (permanent, and three types of temporary employment: fixed-term, temporary employment agency, seasonal work), the following factors were also included as independent variables: employee *skill level* and *contract of choice*. The skill level included several categories of workers, from *unskilled blue-collar* to *management/director*. The authors of this study proved that temporary workers achieved better results in terms of health and organizational behavior, regardless of the temporary contract type, than permanent-contract employees.

The *skill level* analysis showed that both high- and low-skilled temporary workers had better health and more positive work attitudes than permanent workers. With regard to the *contract of choice*, the study results revealed that the highest health and performance indicators were observed among those employees who accepted their type of employment contract, followed by temporary workers, with the permanent workers recording the lowest scores. The quality of the psychological contract proved to be a full or partial mediator between the type of employment contract and the health and performance indicators. The authors concluded that British temporary workers were healthier and performed better at work than permanent workers, partly because the state of their psychological contracts was better. The authors explain such results with a general deterioration in the quality of work performed by permanent workers in the UK, resulting from unsatisfactory levels of employer psychological contract fulfillment noted among permanent-contract employees, namely the falling number of employer contract obligations and frequent cases of psychological contract breach.

It can therefore be assumed that the psychological contract, and particularly its two elements: the number of obligations fulfilled (hereinafter referred to as 'psychological contract fulfillment' (PCF)) and breach, constituting an employee subjective assessment of the quality of employment, would also prove to be an important mediator in relation to selected employment characteristics and the health and performance of Polish employees.

4.2 PRESENT STUDY

Imhof and Andresen (Imhof 2018), based on their review of previous studies on the well-being and health of temporary workers, have concluded that a large variety of different temporary employment forms are summarized under the term temporary work in the existing research. This imposes limits on the comparability and generalizability of empirical findings to date. Differences in job characteristics and job quality in fixed-term employment, temporary agency work, or of on-call employees, leading to divergent effects on well-being, have been lost in previous studies due to the use of the umbrella term *temporary work*.

A further weakness of the research is the very low availability of data that correspond to countries with the highest percentage of temporary employment, such as Poland, Spain, France, or Germany. Moreover, previous studies have shown differences between temporary employment populations in terms of stress, health, work attitudes, and performance (Ray et al. 2017; De Witte and Näswall 2003); however, these studies have failed to account for each country, legal frameworks, the roles played by institutional actors, market structures, and the labor market situation, and these factors in turn shape the characteristics that temporary work has in each national context.

Another criticism that could be leveled against the existing research on the relationship between temporary employment and employee health and performance is the common neglect of other variables such as social status of the type of work performed, or contract volition (e.g. De Cuyper and Witte 2008a; Gracia 2011; Kauhanen and Natti 2015; Lopes and Chambel 2017).

The present study aims to overcome these shortcomings. It focuses on persons remaining in temporary employment in Poland, which for over a decade has been the country with the highest rates of temporary employment in Europe. Moreover, the Eastern European welfare regime in Poland is characterized by a decline in unionization, curbing the advancement of workers' rights and resulting in higher levels of perceived job insecurity (Dixon, Fullerton, and Robertson 2013; Bambra et al. 2014), and low temporary-to-permanent-contract transition rates (Eurofound 2015; Pańków 2015).

In order to apply a more homogenous perspective, the present study has focused on a single type of temporary employment, i.e. fixed-term employment. Moreover, the present research model has introduced other predictors as important variables, apart from temporary vs. permanent employment. Based on Guest and Clinton's (Guest and Clinton 2006) work, these are the contract of choice/volition, and the type of work performed (non-manual vs. manual) reflecting the occupational status of the employee. Work ability as a dimension of a worker's health has also been introduced as a dynamically developing research area in the occupational health field. According to its authors (Ilmarinen and Tuomi 2004), work ability includes subjectively evaluated aspects of mental and physical health, including the type of work performed (physical, mental, and mixed), and identifies diseases diagnosed by a physician. The work ability significantly predicts the length of employee occupational activity, including premature quitting of the labor market (Roelen et al. 2018). Since the diagnostic tool used to measure work ability focuses mainly on the physical health of employees, an additional assessment of mental health has also been introduced into the present research.

Two aspects of psychological contract have been taken into account as important mediators in the relationship between employment volition, type of work/skill level, and health and work performance of temporary and permanent workers, identified by Guest et al. (Guest and Clinton 2006; Guest et al. 2010; Isaksson et al. 2010). This model has been tested in the UK and several European countries and Israel and is used in the current study among the Polish permanent and fixed-term workers (Figure 4.1).

The aim of the present study has therefore been to determine the direct and indirect relationships between the type of employment contract (permanent vs. fixed-term), employment contract volition (voluntary vs. involuntary), the type of work performed (non-manual vs. manual) and health (work ability, mental health), performance (productivity and work engagement), and psychological contract (contract obligations and contract breach). These relations are illustrated in Figure 4.1.

On the basis of the existing research, including national study data, the following study hypotheses have been formulated:

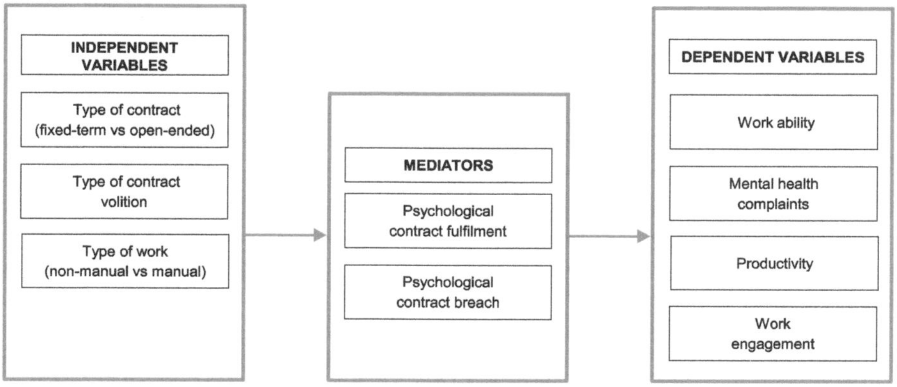

FIGURE 4.1 Conceptual model of relationships between type of contract, volition, type of work, PC, work ability, mental health, productivity, and work engagement.

H1. Fixed-term employment is negatively associated with employee health (mental health and work ability) and performance (work engagement and productivity).

H2. Psychological contract fulfillment is a mediator in the relationship between type of employment contract and employee health (mental and physical) and performance (work engagement and productivity).

H3. Psychological contract breach is a mediator in the relationship between type of employment contract and employee health (mental and physical) and performance (work engagement and productivity).

H4. Employment contract volition is positively associated with employee health (mental and physical) and performance (work engagement and productivity).

H5. Psychological contract fulfillment is a mediator in the relationship between employment contract preference and employee health (mental and physical) and performance (work engagement and productivity).

H6. Psychological contract breach is a mediator in the relationship between employment contract preference and employee health (mental and physical) and performance (work engagement and productivity).

H7. Physical work is negatively associated with employee health (mental and physical) and performance (work engagement and productivity).

H8. Psychological contract fulfillment is a mediator in the relationship between the type of work and employee health (mental and physical) and performance (work engagement and productivity).

H9. Psychological contract breach is a mediator in the relationship between the type of work and employee health (mental and physical) and performance (work engagement and productivity).

4.2.1 Method

4.2.1.1 Study Procedure and Participants

A cross-sectional questionnaire study was carried out. In order to preserve distribution of the fixed-term employment variable comparable to the Polish population, a random-quota sampling method was applied. Twenty-five randomly selected Polish enterprises, located in different regions of the country, from the construction, manufacturing, and retail/service sectors were invited to participate in the study. The selected industry sectors are characterized by the highest rates of fixed-term contracts and type of work diversity (manual, non-manual, partly manual, partly non-manual) in Poland (Lewandowski et al. 2017). Initially, 1,526 employees were invited to participate in the study. The fixed-term employee study sample comprised 920 persons, whereas the permanent employee study sample included 406 persons. The participants were invited to fill out confidential questionnaires, either during working hours or at home. Each participant provided written informed consent prior to the study. The study also obtained an approval of the Ethics Committee under the Declaration of Helsinki.

In total, 1,000 completed questionnaires were collected (response rate = 65%), whereby 700 were answered by fixed-term contract employees and the remaining 300 by permanent employees; 50.2% of the entire study sample were women, 49.8% were men. The mean age of the respondents was 36.046 years (*SD* = 9.70). The average career length was 6.57 years (*SD* = 5.92). Table 4.1 presents descriptive data of the entire study sample, distinguished into fixed-term and permanent employee groups.

4.2.1.2 Questionnaires

Type of employment contract was measured with the question: *what kind of employment contract do you have?* The respondent provided the answer: *fixed-term contract* – coded as '1', or *indefinite-term contract* – coded as '2'.

Contract type volition was measured with a single item: *my present employment contract suits me for the time being*, which was used in the study conducted by De Cuyper and De Witte (De Cuyper and Witte 2008a), exploring contract volition and reasons for accepting fixed-term employment. The respondents indicated: *yes* – coded as '2', or *no* – coded as '1'.

Type of work was measured with the following question: *is your work: (1) mainly psychologically demanding, or (2) mainly physically demanding?* The respondents were asked to select one of the two answers, where the *psychologically demanding* work is referred to as 'intellectual work' – coded as '1', and the *physically demanding* work is referred to as 'physical work' – coded as '2'.

Work ability was assessed using a simplified version of Work Ability Index (WAI) developed by Tuomi et al. (Tuomi 1991) at the University of Wupertal with the agreement and cooperation of the main authors of the original scale (Hasselhorn 2008). The scale comprises seven items.

The respondent evaluates: (1) current work ability compared with lifetime best (range: 0–10), (2) work ability in relation to the demands of the job (range 2–10), (3) number of current diseases diagnosed by a physician (range 1–7), (4) estimated work impairment due to diseases (range 1–6), (5) sick leave during the past year (12 months) (range 1–5), (6) own prognosis of work ability two years from now (range 1–7), (7) mental resources (range 1–4). A high aggregate score indicates a high work ability. The questionnaire reliability was Cronbach's $\alpha = 0.82$.

Mental health was assessed using the Polish version (Makowska and Merecz 2001) of *The General Health Questionnaire-28* (GHQ-28) developed by Goldberg and Williams (Goldberg and Williams 1991). The questionnaire constitutes one of the most popular mental health assessment tools. The GHQ-28 consists of 28 items comprising 4 diagnosis subscales: somatic complaints, anxiety and insomnia, social dysfunction, and depression. Participants were asked to respond to each item on a four-point scale, ranging from *not at all* to *much better than usual*, or from *better than usual* to *much worse than usual*. The answers scored respectively: 0-0-1-1. The aggregate score is the sum of points scored for all questionnaire items, where the maximum possible score is 28 points. The higher the score, the higher the level of mental health complaints, i.e. the poorer the mental health. In the conducted study the mental health measurements obtained the following, satisfactory reliability indexes: somatic symptoms (0.75), anxiety and insomnia (0.86), functional disorders (0.76), and depression (0.87).

Performance was assessed with the *Perceived Group Performance scale* (Conger et al. 2000) which includes five items describing employee perception of his/her

TABLE 4.1
Fixed-Term-Contract and Permanent-Contract Employee Groups' Characteristics

		Fixed-Term Contract		Open-Ended Contract		Total	
	Variables	*N*	%	*N*	%	*N*	%
Gender	Female	351	50.2	146	48.8	498	49.8
	Male	349	49.8	154	51.2	502	50.2
	Total	700	100	300	100	1000	100
Marital status	Married	325	46.5	191	63.5	516	51.6
	Cohabitating	114	15.5	36	12.0	150	15.0
	Single	192	27.5	40	13.3	232	23.2
	Divorced	35	5.0	23	7.6	58	5.8
	In remarriage	16	2.3	7	2.3	23	2.3
	In separation	8	1.1	1	0.3	9	0.9
	Widower/widow	7	1.0	3	1.0	10	1.0
	Total	697	99.7	300	100	998	99.8
Age category	18–29	246	35.2	48	15.9	294	51.1
	30–39	259	37.1	109	36.2	368	73.3
	40–49	134	19.2	87	28.9	221	48.1
	50–64	56	8.0	55	18.3	111	26.3
	Total	695	99.4	299	99.3	942	99.4
Education	Grammar school or lower	11	1.6	2	0.7	13	1.3
	Lower vocational	83	11.9	46	15.3	129	12.9
	Upper vocational	181	25.9	51	16.9	232	23.2
	Secondary school	105	15.0	30	10.0	135	13.5
	Post-secondary	147	21.0	55	18.3	202	20.2
	Higher	169	24.2	117	38.9	286	28.6
	Total	696	99.6	300	100	997	99.7
Sector	Private	121	82.4	90	70.1	211	21.1
	Public	576	17.3	211	29.9	787	78.7
	Total	697	99.7	300	100.0	998	99.8
Occupation	Drivers	17	2.9	24	8.0	41	4.1
	Specialists	87	14.3	88	29.2	175	17.5
	Technical and middle staff	71	9.6	30	10.0	101	10.1
	Office workers	106	15.2	36	12.0	142	14.2
	Personal services and retail workers	108	15.5	41	13.6	149	14.9
	Industry processing workers and craftsmen	143	20.5	33	11.0	176	17.6

(Continued)

TABLE 4.1 (CONTINUED)
Fixed-Term-Contract and Permanent-Contract Employee Groups' Characteristics

		Fixed-Term Contract		Open-Ended Contract		Total	
	Variables	*N*	%	*N*	%	*N*	%
	Machinery assemblers and operators	41	5.8	29	9.6	70	7.0
	Low-skilled labor workers	124	17.7	20	6.6	144	24.4
	Total	697	99.7	300	100	998	99.8
Type of work	Non-manual	256	36.2	208	68.5	464	46.4
	Manual	444	63.8	92	31.5	536	53.6
	Total	699	100	300	100	1000	100
Contract volition	Involuntary	304	43.3	20	6.2	323	32.3
	Voluntary	396	56.7	280	93.4	677	67.7
	Total	700	100	300	100	1000	100
Average job tenure		$M = 3.6$	$SD = 0.8$	$M = 9.2$	$SD = 2.4$	$M = 6.9$	$SD = 2.7$

team performance. The respondent specifies the perceived performance on a six-point scale, where 6 is *strongly agree*, and 1 is *strongly disagree*. A high aggregated score translated into a high performance. A high study reliability Cronbach's α = 0.91 was noted.

Work engagement was measured using an abbreviated three-item version of the *Utrecht Work Engagement Scale* (*UWES*) developed by Schaufeli and Bakker (2003) in the Polish translation by Szabowska-Walaszczyk et al. (2011). The following items were characterized by the strongest correlation with the entire UWES questionnaire: *I feel strong and full of energy at work*, *I am full of enthusiasm for my work*, and *I feel happy when I work intensively*. The respondents provided answers on a six-point scale, where 0 was *never*, and 6 – *always/each day*. A high aggregate score reflected a high level of work engagement. A high study reliability Cronbach's α = 0.90 was noted.

Psychological contract fulfillment (PCF) and ***psychological contract breach (PCB)*** were measured with the *Psychological Contract* diagnosis test, developed by Coyle-Shapiro and Kessler (2000). The questionnaire comprises 2, 14-item subscales assessing 3 types of employer obligations: financial (e.g. *pay adequate to my duties*, *additional and fair benefits*, *compared to similar position employees in other organizations*), psychosocial (e.g. *social support*, *opportunity to participate in decision-making*), and professional development opportunities (*career prospects*, *training that allows me to do my job well*), whereby the first obligation is of a transactional, and the remaining two obligations of a relational nature (Rousseau 1995; Rousseau and Tijoriwala 1998). The respondents indicated on a five-point Likert scale (1 – *not at all*, 5 – *to a very great extent*) the extent to which they believed their employer was obliged to fulfill his/her obligations – corresponding to the *Perceived employer obligations* (expected contract fulfillment) subscale; or the extent to which the employer has actually fulfilled his/her inducements – corresponding to the *Perceived employer inducements* (actual contract fulfillment) subscale. The higher the aggregate subscale score, the higher the psychological contract fulfillment.

Psychological contract breach (PCB), defined as the degree of discrepancy between the expected contract fulfillment and the actual contract fulfillment, was measured by the difference between the *Perceived employer obligations* subscale score, and the *Perceived employer inducements* subscale score. The more positive the score, the higher the PCB. Conversely, the more negative the score, the lower the PCB. A high study reliability Cronbach's $\alpha = 0.93$ was noted for both subscales.

4.2.1.3 Statistical Analyses

The analyses of the study results were carried out in two steps. Firstly, descriptive statistics of the two employee groups, fixed-term- and permanent-contract employees, were completed. The mean scores of permanent employment contract respondents were compared with the fixed-term employment contract respondents' results in relation to the analyzed dependent variables: work ability, four mental health components (somatic complaints, anxiety and insomnia, social dysfunction, and depression), performance, and work engagement. A correlation analysis of all variables was also conducted in the first step.

In the second step, the main study analyses testing the study model were carried out using the Mplus 7.3 software. Structural equation modeling and the Maximum Likelihood Robust (MLR) method were used to verify the continuous, non-normal variable distribution model. A multiple mediation model (Preacher et al. 2006) with two mediators (psychological contract and psychological contract breach) was used in the relationship between three predictors: (1) type of employment contract (fixed-term vs. permanent), (2) employment contract volition (voluntary vs. involuntary), (3) type of work (manual vs. non-manual), and four dependent variables: (1) work ability, (2) mental health complaints, (3) performance, (4) work engagement.

The model data fit estimation was performed according to fit indices developed by Kline (Kline 2005). The most widely used data fit indices were applied, i.e. root mean square error approximation (*RMSEA*), standardized root mean square residual (*SRMR*), comparative fit index (*CFI*), and the Tucker–Lewis index (*TLI*), as well as the χ^2 general fit index and its probability.

A mediation analysis was conducted in order to verify the mediating role of psychological contract and psychological contract breach in the relationship between employment characteristics and employee health and performance. The analysis was performed using the bootstrapping method (Preacher and Hayes 2008) with a random sampling of 5,000 bootstrap samples. The mediation analysis enabled a more complex structure of the model, in which three independent variables, acting as predictors (type of employment contract, contract volition, type of work), were associated with the dependent variables (work ability, mental health complaints, performance, work engagement) by other variables acting as mediators (psychological contract and psychological contract breach). The mediation effect occurs when the mediating variable mitigates the relationship between the independent variable and dependent variable. The IBM SPSS Statistics PROCESS procedure version 22.0 was used to perform the calculations.

4.3 RESULTS

4.3.1 Preliminary Analyses

Table 4.1 presents data on gender, age category, marital status, education, sector (public vs. private), position held, type of job, type of employment, contract volition, and career length in the entire study sample, and in distinct fixed-term and permanent employee groups.

Student's *t* test was used to compare the two employee groups and revealed that persons employed under fixed-term contract were significantly younger in comparison with permanent-contract employees (Table 4.2). Moreover, the psychological contracts of fixed-term employees were significantly narrower and more frequently breached than those of permanent employees (Table 4.2). In addition, the analysis revealed further significant differences between these groups. The results showed that persons employed on a fixed-term basis were significantly more likely to perform manual rather than non-manual work (U = 713.00 $p < 0.001$), were predominantly industrial workers and performed simple jobs (U = 858.00, $p < 0.001$), and their type of employment was more frequently involuntary ($\chi^2 = 224.43$, $p < 0.000$) compared to permanent employees.

The correlation analysis of the study variables revealed significant correlations, mainly between work ability, performance, work engagement, psychological contract, and psychological contract breach and other variables (Table 4.3). Significant correlations with selected variables also concerned gender, age, type of work, and type of employment (Table 4.3).

TABLE 4.2
Fixed-Term and Open-Ended Contract Employee Groups' Differences – *t Student Test* Results

	Fixed-Term Contract		Open-Ended Contract			
	M	*SD*	*M*	*SD*	*t*	*p*
Age	34.44	9.27	39.77	9.68	**–8.21**	**0.00**
Work ability	40.38	5.37	40.64	5.01	–0.71	0.48
Mental health complaints	47.15	9.28	47.31	8.47	–0.303	0.77
Performance	17.90	4.05	18.34	3.96	–1.62	0.11
Work engagement	13.54	4.00	13.83	4.04	–1.06	0.29
PCF	50.91	11.42	54.43	11.86	**–4.41**	**0.00**
PCB	10.93	14.68	7.83	13.49	**3.23**	**0.00**

TABLE 4.3
Correlation Analysis (Pearson's r) between Variables

	1. Gender	2. Age	3 Type of Work	4. Type of Employment	5. Contract Volition	6. Work Ability	7. Mental Health Complaints	8. Performance	9. Work Engagement	10. PCF	11. PCB
1.	1	–0.071*	–0.139**	0.013	–0.025	–0.110**	0.191**	0.075*	0.026	0.014	0.038
2.	–0.071*	1	–0.030	0.252**	0.106**	–0.276**	0.132**	–0.053	–0.043	0.009	–0.110**
3.	–0.139**	–0.030	1	–0.159**	–0.088**	–0.201**	0.015	–0.206**	–0.216**	–0.096**	–0.061
4.	0.013	0.252**	–0.159**	1	0.360**	0.023	0.010	0.051	0.034	0.138**	–0.099**
5.	–0.025	0.106**	–0.088**	0.360**	1	0.092**	–0.091**	0.075*	0.087**	0.101**	–0.161**
6.	–0.110**	–0.276**	–0.201**	0.023	0.092**	1	–0.434**	0.375**	0.552**	0.258**	0.044
7.	0.191**	0.132**	0.015	0.010	–0.091**	–0.434**	1	–0.167**	–0.283**	–0.204**	0.055
8.	0.075*	–0.053	–0.206**	0.051	0.075*	0.375**	–0.167**	1	0.449**	0.366**	0.288**
9.	0.026	–0.043	–0.216**	0.034	0.087**	0.552**	–0.283**	0.449**	1	0.245**	–0.049
10.	0.014	0.009	–0.096**	0.138**	0.101**	0.258**	–0.204**	0.366**	0.245**	1	–0.291**
11.	0.038	–0.110**	–0.061	–0.099**	–0.161**	0.044	0.055	0.288**	–0.049	–0.291**	1

* $p < 0.05$ (two-tailed).
**$p < 0.01$ (two-tailed).

4.3.2 Age, Gender, and Employee Health and Performance, and Psychological Contract

The obtained results have revealed that older persons ($\beta = -0.36$, $SE = 0.03$, $p < 0.001$) and women ($\beta = -0.18$, $SE = 0.03$, $p < 0.001$) have significantly poorer work ability. Older employees ($\beta = 0.19$, $SE = 0.04$, $p < 0.001$) and women ($\beta = 0.23$, $SE = 0.04$, $p < 0.001$) also have lower mental health indicators, i.e. they report more mental health complaints. Older persons have also declared a poorer (narrower) psychological contract ($\beta = -0.19$, $SE = 0.04$, $p < 0.001$) and are characterized by lower performance levels ($\beta = -0.10$, $SE = 0.03$, $p < 0.01$) compared to younger workers.

4.3.3 Main Study Analyses – Hypotheses Testing

Next, the model (as presented in Figure 4.2) data fit was verified. The results revealed that the two-mediator (the psychological contract fulfillment (PCF) and psychological contract breach (PCB)) model had a good data fit, $\chi^2 = 132.94$, $df = 57$, $p < 0.001$. The analysis of the other indicators also revealed a good data fit of the model: $CFI = 0.96$, $TLI = 0.94$, $SRMR = 0.03$, $RMSEA = 0.04$, 90% CI (0.03, 0.05), $p > 0.05$ ($p = 0.87$).

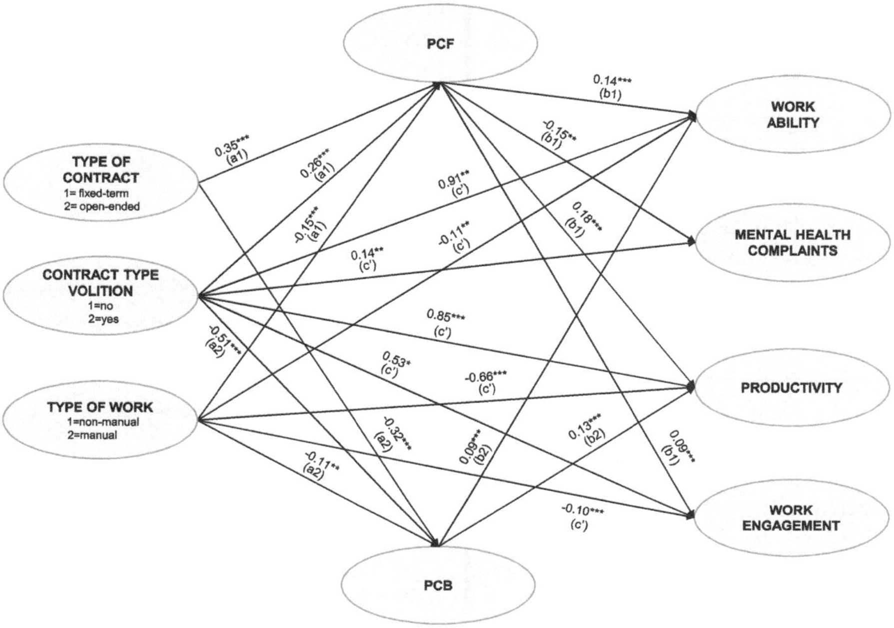

FIGURE 4.2 Empirical model of relationships between type of contract, volition, type of work, PC, work ability, mental health, productivity, and work engagement.

4.3.4 Type of Employment and Employee Health (Work Ability, Mental Health), Performance (Productivity, Work Engagement), and Psychological Contract

The path coefficient significance analysis has revealed a lack of any significant direct relationships between the type of employment contract and work ability, mental health, work engagement, and performance, which means that Hypothesis 1 has not been confirmed (Figure 4.2).

However, several indirect relationships between the type of employment and these dependent variables have been observed where the psychological contract fulfillment and the psychological contract breach have proved to be important mediators. Accordingly, the psychological contract fulfillment was a total mediator in the relationship between the type of employment and work ability ($\beta_{indirect}$ = 0.46, 95%, *CI*: 0.23, 0.71), mental health ($\beta_{indirect}$ = –0.56, 95%, *CI*: –0.89, –0.28), performance ($\beta_{indirect}$ = 0.62 95%, CI: 0.34, 0.90), and work engagement ($\beta_{indirect}$ = 0.30, 95%, *CI*: 0.16, 0.47). This means that permanent-contract employees have more psychological contract promises fulfilled and, as a result, higher work ability, better mental health, and higher performance and work engagement levels than fixed-term contract employees. Thus, Hypothesis 2 has been fully confirmed.

Also, the psychological contract breach was found to be a total mediator between the type of employment and work ability ($\beta_{indirect}$ = –0.19, 95%, *CI*: –0.27, –0.10), and performance ($\beta_{indirect}$ = –0.40, 95%, *CI*: –0.63, –0.16). This means that fixed-term employees report a higher degree of psychological contract breach and, as a result, have a lower work ability and performance levels than permanent workers. Hypothesis 3 has therefore been confirmed with regard to work ability and performance.

4.3.5 Employment Contract Volition and Employee Health, Performance, and Psychological Contract

The results of the study analyses have revealed that the contract volition is directly associated with work ability ($\beta = 0.91$, $SE = 0.34$, $p < 0.01$), negatively associated with mental health complaints ($\beta = -0.14$, $SE = 0.60$, $p < 0.01$), and positively associated with performance ($\beta = 0.85$, $SE = 0.30$, $p < 0.000$) and work engagement ($\beta = 0.53$, $SE = 0.28$, $p < 0.03$). This means that those employees who have a contract of choice have a higher work ability, performance, and work engagement, as well as better mental health than those who do not have a contract of choice (Figure 4.2). Hypothesis 4 has therefore been fully confirmed.

The psychological contract fulfillment has also proved to be an important partial mediator in the relationship between employment contract volition and work ability ($\beta_{indirect}$ = 0.31, 95%, *CI*: 0.99, 0.55), mental health complaints ($\beta_{indirect}$ = –0.38, 95%, *CI*: –0.68, –0.14), performance ($\beta_{indirect}$ = 0.43, 95%, *CI*: 0.16, 0.69), and work engagement ($\beta_{indirect}$ = 0.22, 95%, *CI*: 0.10, 0.38). Hence, persons who have a contract of choice have a higher work ability, better mental health, and higher performance and work engagement levels, in part owing to a greater fulfillment of psychological contract employer obligations. Hypothesis 5 has therefore been fully confirmed.

Psychological contract breach has proved to be a significant, partial mediator in the relationship between the employment contract volition and work ability ($\beta_{indirect}$ = −0.25, 95%, *CI*: −0.41, −0.11), and performance ($\beta_{indirect}$ = −0.13, 95%, *CI*: −0.21, −0.39). This implies that persons who do not have a contract of choice have a poorer work ability and performance, in part owing to more frequent psychological contract breach than those who have a contract of choice.

Hypothesis 6 has therefore been confirmed in relation to two out of four dependent variables, i.e. work ability and performance.

4.3.6 Type of Work and Employee Health, Performance, and Psychological Contract

The type of work has been found directly negatively associated with work ability ($\beta = -0.11$, $SE = 0.20$ $p < 0.000$), performance ($\beta = -0.66$, $SE = 0.13$, $p < 0.000$), and work engagement ($\beta = -0.10$, $SE = 0.15$, $p < 0.000$). This means that workers performing manual work have a poorer work ability, performance, and work engagement than high-skilled workers. This confirms Hypothesis 7 for all dependent variables but mental health.

Indirect relationships between the type of work and work ability, mental health, and performance have also been revealed, in which psychological contract fulfillment has been an important partial mediator. An indirect negative relationship between type of work and work ability ($\beta_{indirect}$ = −0.22, 95%, *CI*: −0.31, −0.16), and performance ($\beta_{indirect}$ = −0.24, 95%, *CI:* −0.40, −0.10), and a positive relationship between type of work and mental health complaints ($\beta_{indirect}$ = 0.24, 95%, *CI*: 0.10, 0.39) have also been identified. In the latter case, a full mediation was observed. This suggests that low-skilled workers have lower work ability, and poorer work performance and work engagement, partially because they have fewer psychological contract promises fulfilled. Accordingly, these employees report more mental health complaints. Hypothesis 8 has therefore been confirmed with regard to all the dependent variables.

Also, the psychological contract breach has proved to be a significant partial mediator in the relationship between the type of work and work ability ($\beta_{indirect}$ = −0.14, 95%, *CI*: −0.10, −0.25), and performance ($\beta_{indirect}$ = −0.19, 95%, *CI*: −0.25, −0.10). This means that low-skilled workers have lower work ability and poorer performance levels, partially because they experience psychological contract breach more frequently than high-skilled workers. Hypothesis 9 has therefore been confirmed with regard to work ability and performance.

4.4 DISCUSSION

The present study results indicate that the model developed by the British researchers has proved to fit the Polish data; however, in most cases, different relationships have been revealed. An important finding of the present research, unlike the British study, has revealed no direct relationship between the type of employment and employee health and performance. However, the quality of the psychological contract, and in particular the number of the employer's promises and their fulfillment, has proved to be a mechanism that significantly determines whether temporary workers enjoy good physical and mental health and are productive and engaged at work. Failure to include this key variable in previous research may be the reason for the inconsistent findings obtained so far. Some of these findings have supported the relationship between temporary employment and poor mental and physical health of employees (Martens et al. 1999; Rodrigues 2002; Virtanen et al. 2002; Virtanen et al. 2003a; Virtanen et al. 2005; Vives et al. 2013; Jang et al. 2015; Canivet et al. 2016; Vives et al. 2017; Kim et al. 2016; Kachi et al. 2014), whereas no clear differences between the two employee groups have been identified by other authors (Virtanen 2001; Bardasi and Francesconi 2004; Artazcoz et al. 2005). Some findings have also reported that temporary workers enjoy better well-being than permanent employees (Virtanen et al. 2001; Goudswaard and Andries 2002; De Cuyper 2006a).

The present study has shown that if the quality of temporary workers' psychological contract is inferior to that of permanent workers, then there are also negative effects on the physical and mental health of temporary employees and their work performance in terms of reduced work engagement and productivity. The number of psychological contract obligations has proved to be greater for permanent workers than for fixed-term employees. This result has been supported by the other national data (Lewandowski et al. 2017) and the data from Dutch studies which have revealed that temporary workers are more likely to have an imbalanced employment contract in terms of employer and employee obligations, to the detriment of the latter (De Jong et al. 2009; De Cuyper and Witte 2007a). De Witte and Näswall (2003) examined the opinions expressed by employers and HR managers on the working conditions/contracts of temporary workers. The researchers found that although most of the surveyed organizations declared an equal treatment of both temporary and permanent workers, the HR managers ultimately admitted that their organizations offered more fringe benefits to permanent staff than to temporary-contract employees. These findings have also been supported by non-European studies. Indian (Bhandari and Hesmati 2006) and Australian (Wooden 2004) researchers proved that temporary agency workers were paid lower wages than permanent workers in these countries. Data on the U.S. and Australian labor markets have also shown significantly smaller fringe benefits for temporary workers (Nollen 1996; Kalleberg 2011). Zeytinoglu and Cooke (2005) have also noticed lower promotion opportunities for temporary workers. Several other studies have also revealed that temporary employees have significantly lower chances of further education and participation in employer-paid training (Aronsson et al. 2002; Forrier and Sels 2003; Connelly and Gallagher 2006). This has also been reflected in the latest OECD report (OECD 2019). Similarly, Polish studies on the working conditions of fixed-term employees are consistent with the earlier findings, and prove that these persons usually work in companies offering lower wages (even by several percentage points) to fixed-term workers than permanent employees (Goraus and Lewandowski 2016; Pańków 2015).

In turn, the present study results, which demonstrate that the lower job quality of persons employed on fixed-term contracts is conducive to poorer health, have been supported by a study conducted by Bernhard-Oettelet et al. (Bernhard-Oettelet 2005). The researchers revealed that there were specific work aspects (job insecurity, job control, and job demands) that proved to be more important predictors of worker health than the type of employment contract. Adverse health outcomes were observed among workers reporting high job demands and low job control in task performance. The findings thus prompted the authors to conclude that subjectively assessed working conditions were a stronger predictor of worker health and well-being than objective conditions, such as formal employment status.

Similar conclusions have been drawn in the present study, which has shown no direct relationship between the type of employment contract and work engagement, as well as the performance of employees. As mentioned earlier, previous research has produced inconsistent results so far, showing that temporary employees are characterized by lower levels of work engagement (De Witte and Näswall 2003; Connelly and Gallagher 2006; De Cuyper et al. 2008b), or equally proving that a lower work engagement has been observed among permanent workers (De Cuyper and De Witte 2006b; De Cuyper and De Witte 2007a; De Witte and Näswall 2003; McDonald and Makin 2000; Chambel and Castanheira 2006). Such a discrepancy – similar to the relationship between the type of employment and employee health – may be due to the conceptual heterogeneity applied in the analyses of temporary employment, as well as the diversity of work cultures considered in those studies. As in the case of the research on employee health, the obtained results may depend on the type of temporary employment, and the local country settings that the research has been focused on.

The current study demonstrates that there is an indirect relationship between the type of employment contract and physical and mental health, performance, and work engagement of employees, whereby the psychological contract plays a mediating role, proving to be an important total mediator of all employment relationships and employee health and performance. This result suggests that high job quality and good working conditions are a crucial driver of strong employee performance. The evidence has been supported by Fulmer et al. (Fulmer 2003) who analyzed the top 100 companies in the U.S. in terms of job quality and enterprise efficiency, and concluded that it was precisely the high job quality that produced the greatest efficiency rates.

Recently, Dineen and Allen (Dineen 2016) have re-examined the top 100 companies with the highest employment standards and have concluded that investing in multiple HR activities addressed to employees generates employer benefits in terms of lower staff turnover, and the potential of attracting job seekers. Several other studies also suggest that a positive employee perception of the human resource management at the workplace can lead to increased levels of work engagement and job satisfaction, which in turn foster organizational citizenship behavior and stronger employee performance (Allen et al. 2003; Kuvaas 2008; Snape and Redman 2010). Various supportive measures addressed to employees by employers are also very important in the context of the work engagement of temporary workers. Chambel and Sobral (Chambel 2011) proved that temporary employees were significantly more engaged at work, if the employer had provided training opportunities, improving individual employability and increasing the likelihood of finding a new job. Kinnunen et al. (Kinnunen 2011), who obtained similar results in a group of Finnish employees, concluded that the employer fear to offer professional training to short-term employees is unjustified, because the evidence proves that such an investment yields significant profits in terms of stronger work engagement and performance of temporary workers. Employer supportive actions, aiming at increased employee participation in decision making and improvement of communication processes, remarkably strengthen employee confidence in their own employability, even under conditions of company restructuring and a potential threat of losing their job (Kammeyer-Mueller and Liao 2006).

Correspondingly, the novelty of the present study has been the finding revealing that a high-quality psychological contract is not only significantly associated with the performance but also with the work ability of employees. So far the research has not provided sufficient evidence proving that employer supportive actions (apart from health schemes) are conducive to the improved physical health of employees, including both the WAI-diagnosed health status and the predicted work ability, which is considered an important predictor of employee participation in the labor market.

Another predictive factor of health and work attitudes of persons employed under temporary contracts indicated in the published literature has been the compatibility of the employment contract with employee contract preferences. The present study evidence proves that persons who are in involuntary employment (and these are significantly more often temporary workers) have worse physical and mental health indicators, and are characterized by lower performance and work engagement. A similar result was obtained by Guest and Clinton (Guest and Clinton 2006) who, in the aforementioned study, revealed that a voluntary type of employment was directly, positively associated with improved health and higher performance of employees. In the present study, the psychological contract obligations – significantly greater for those who have a contract of choice – have proved to be a partial mediator of these relationships. Similarly, Kauhanen and Natti (Kauhanen 2015) analyzed the Finnish annual reports on the quality of working conditions in 1997, 2003, and 2008, and observed that employees with a temporary contract of choice rated higher such working conditions as educational opportunities, employer-paid training, career development opportunities, and job control, than those workers who had an involuntary temporary work contract.

The present research has also revealed that persons with an involuntary work contract have lower work ability and poorer work performance, partly due to more frequent experiences of psychological contract breach, compared to persons with a contract of choice. The cited study of Guest and Clinton (Guest and Clinton 2006) also showed that British temporary workers for whom this type of employment was not a preferred choice scored significantly lower in six out of seven mental health indicators, as well as in five out of eight work attitude indicators surveyed. The mediating factor in these relationships proved to be – just as in the present study – psychological contract breach. De Jong and Schalk (2010) explain a similar finding obtained in a study on Dutch temporary workers. The involuntary nature of the temporary contracts was found by the researchers much more conducive to psychological contract breach than in the case of temporary contracts of choice.

Furthermore, the present study has also revealed, in Poland, persons employed on fixed-term contracts are younger, perform physical rather than intellectual work, and the temporary nature of their employment is more often incompatible with their contract preferences in comparison with permanent employees. Nunez and

Livanos (Nunez 2011) surveyed the population of European young workers by type of employment and identified two broad categories of young temporary workers: (1) those who had a fixed-term contract of choice as a way of gaining wider professional experience, where this type of employment contract was perceived as a stepping-stone, allowing greater flexibility – a common practice in Nordic and continental countries; (2) persons who took up an involuntary temporary employment and were not satisfied with it, as they would have preferred a permanent job contract which was not yet available – characteristic for Mediterranean countries (Greece, Italy, Portugal). The obtained results revealed that Polish fixed-term workers corresponded to the Mediterranean, rather than the Nordic or continental group, since 50% declared that they could only find temporary employment, 28% perceived it as a way of transitioning to a permanent contract, and only 12% engaged in fixed-term employment as a matter of personal choice and flexibility. However, compared to the years 2012–2014, when fixed-term employment contracts in Poland were predominantly involuntary (almost 70% of temporary workers could not find permanent employment at that time (Eurofound 2015; Pańków 2015)), the condition of temporary workers has now slightly improved. However, despite the decrease in the volume of involuntary temporary contracts in Poland, further efforts should be made at a greater reduction of such contracts. The current study results indicate that an involuntary type of employment can lead to poorer physical (lower work ability) and mental health of workers. Hiring such employees does not pay off either, as employers cannot count on the higher performance of workers who are not satisfied with the type of their employment.

Another predictor of employee health and performance analyzed in the present study has been the type of work performed. The obtained results suggest that low-skilled workers have poorer work ability, and are also characterized by lower performance, i.e. lower productivity and work engagement than intellectual employees. The poorer health and work ability of low-skilled workers has been largely supported by previous research (Vahtera et al. 1999; Lahelma et al. 2012; Aittomäki et al. 2003; Thorsen et al. 2013; Hu et al. 2016; Schouten et al. 2015; Oliv et al. 2017; Tonnon et al. 2019). The lower work ability of physical workers compared to persons performing intellectual work has also been confirmed in Polish studies (Bugajska et al. 2011). Several of the aforementioned reports reveal that temporary work also poses a risk to mental health, but only in relation to physical workers (Han et al. 2017; Hammarström et al. 2011).

The novelty of the present study has been the result showing that persons performing manual work are also characterized by poorer work performance, i.e. productivity and work engagement, and that one of the mechanisms of these relationships is the employee-perceived low quality of psychological contract.

Much research has confirmed that employer failure to fulfil the psychological contract is associated with a range of counterproductive employee behaviors. Workers who experience a psychological contract breach quit jobs more often (Raja et al. 2004; Bloome et al. 2010), record lower levels of work engagement and organizational citizenship behavior (Cassar and Briner 2011; Turnley and Feldman 1999; Uen et al. 2009), are weaker team-players (Kiewitz et al. 2009), and score lower on innovativeness (Ng 2010). The evidence has also proved that cynical attitudes and counterproductive behaviors among employees rise when the staff regularly experience psychological contract breach (Bordia et al. 2008), and the overall organizational culture is poor (Conway and Briner 2005). The cited research results have been analyzed in line with the theory of social exchange (Blau 1964) and its guiding principle of reciprocity. De Cuyper, et al. (De Cuyper et al. 2011) have rephrased this principle into a recommendation for managers hiring temporary staff: *treat workers well to reap the benefits in terms of loyalty and productive behavior.* The present study results show that physical workers have significantly reduced psychological contract obligations comparing to white-collar workers, and are more likely to experience psychological contract breach, filling a significant gap in the research on temporary workers. The finding has only indirectly been confirmed by Duchaine et al. (Duchaine et al. 2017), who, on the basis of a survey conducted among Canadian workers, concluded that the lower the educational level of workers, the poorer their psychosocial working conditions and the higher their levels of work-related stress. Similarly, a French study (Niedhammer et al. 2016) revealed that the higher the professional position of an employee, the higher the job security, the fairer the rewards for work, the greater the professional development opportunities, the higher the level of job control, the more convenient the working time, the stronger the social support at work, and the lower the exposure to workplace bullying and likelihood of depressive disorders. These findings have also been supported by the results of a large European cohort study of 33,443 workers (Schulte et al. 2015).

The present research has also proved that poor working conditions, i.e. reduced psychological contract obligations and a high level of psychological contract breach observed among physical workers, constitute a significant mechanism of not only deteriorating mental health, but also of reduced work ability. This has been confirmed by longitudinal studies showing that prolonged low-quality employment, rightly referred to as precarious employment, is conducive to deteriorating mental, and also physical health of workers (Benach et al. 2014; Pirani and Salvini 2015; Van Aerden et al. 2015; Van Aerden et al. 2016).

This result suggests that it is more difficult for manual than intellectual workers to negotiate the psychological contract with their employer. This may be due to the lower social and psychological capital of persons employed in manual jobs. Paradoxically, such workers are also less likely to increase their human capital because of the poorer access to skill development training, compared to persons performing more complex tasks (Chłoń-Domińczak and 2015; Eurofound 2017), which, in turn, leaves them trapped in the vicious circle of poor-quality employment.

4.5 CONCLUSIONS

While a direct relationship between the type of employment contract and employee health and performance has not been proved, the evidence suggests that both the health and work attitude of fixed-term employees deteriorate when the job quality, defined in the present study as the quality of the psychological contract between an employer and an employee, is poor, meaning that only very few employer financial and relational obligations have been fulfilled. The current study reveals that this is the mechanism of *precarious employment.* The study has also proved that, in Poland, precarious employment is directly associated with an involuntary nature of the employment contract type, typical for persons employed as low-skilled workers. Despite the reduced number of involuntary temporary contracts in Poland, a further effort should be made aiming at curbing this practice. As the current study results have shown, an involuntary employment type may lead to a deterioration in the physical (lower work ability) and mental health and work performance of persons in involuntary temporary employment.

Moreover, efforts should be made at ensuring that psychological contracts of workers performing manual work reflect the psychological contract obligations guaranteed to high-skilled workers. Low-skilled workers, due to lower social and psychological capital, are less likely to negotiate their psychological contract with their employer than persons performing non-manual work. Poor-quality psychological contracts, and thus the limited access to training, for low-skilled workers constitute a more severe obstacle to breaking out of the precarious employment trap, than for persons in higher social status jobs, including those who use temporary employment contract of choice as a stepping stone to permanent employment.

The obtained results clearly show that it is the quality of employment, rather than the job security, that needs to be fostered to ensure employee health. Perhaps a crucial role is played by the access to skill development training, which constitutes one of the key psychological contract employer obligations, as it strengthens employee human capital, and thus, employability, enabling socially disadvantaged workers to escape the trap of precarious or involuntary employment. These results also prove that ensuring decent working conditions is also a prerequisite for enterprise efficiency and competitiveness, which can only be achieved by increasing productivity and the work engagement of employees. The study results also confirm that social relations are governed by the principle of reciprocity, which benefits both employees and employers alike.

4.6 LIMITATIONS

Given the cross-sectional character of the study, it is difficult to predict with certainty the direction of the revealed associations, i.e. whether fixed-term employment is detrimental to employee health and performance, or if persons with poorer general health and/or performance are more likely to fall into the fixed-term employment trap. However, the research has been largely in favor of the former trend. This has been particularly reflected in the panel studies on the relationship between the type of employment and health, such as the Korean (Kim et al. 2008), Japanese (Kachi et al. 2014), Italian (Pirani and Salvini 2015; Pirani 2017; Moscone et al. 2016), and American (Quesnel-Vallée et al. 2010) panel studies. Therefore, the present study results should be further verified against future longitudinal studies conducted in Poland. For instance, it is quite conceivable that the negative consequences of fixed-term employment may arise when employees stay in fixed-term employment for a longer period than the 3.6 years declared in the present study (Table 4.1).

Nor does the cross-sectional character of the study make it possible to determine whether involuntary fixed-term contracts are detrimental to mental health, or if individuals who enjoy greater well-being (higher self-esteem, self-efficacy, optimistic attitude) are more inclined to aim for employment that meets their preferences, than persons with lower well-being indicators. The research in the field has produced inconsistent results so far (Dawson et al. 2005; Kauhanen and Natti 2015), hence only longitudinal research would allow strong conclusions to be drawn.

Equally, the obtained results should be validated in studies accounting for both the subjective assessment of job quality, such as the psychological contract factor, and the objective data on salary levels, fringe benefits, and availability of professional training, as well as other HR practices designed to promote employee well-being (Guest 2017).

REFERENCES

Aittomäki, A., E. Lahelma, and E. Roos. 2003. Work conditions and socioeconomic inequalities in work ability. *Scand J Work Environ Health* 29(2):159–165.

Alavinia, S. M., A. G. de Boer, J. C. van Duivenbooden, M. H. Frings-Dresen, and A.Burdorf. 2008. Determinants of work ability and its predictive value for disability. *Occup Medic* 59(1):32–37.

Alcover, C. M., R. Rico, W. H. Turnley, and M. C. Bolino. 2017. Understanding the changing nature of psychological contracts in 21st century organizations: A multiple-foci exchange relationships approach and proposed framework. *Organiz Psychol Rev* 7(1):4–35.

Allen, B. 2011. The role of professional identity commitment in understanding the relationship between casual employment and perceptions of career success. *Career Develop Int* 16(2):195–216.

Allen, D. G., L. M. Shore, and R. W. Griffith. 2003. The role of perceived organizational support and supportive human resource practices in the turnover process. *J Manag* 29(1):99–118.

Aronsson, G., and S. Göransson. 1999. Permanent employment but not in a preferred occupation: Psychological and medical aspects, research implications. *J Occup Health Psychol* 4(2):152–163.

Aronsson, G., K. Gustafsson, and M. Dallner. 2002. Work environment and health in different types of temporary jobs. *Eur J Work Organiz Psychol* 11(2):151–175.

Artazcoz, L., J. Benach, C. Borrell, and I. Cortès. 2005. Social inequalities in the impact of flexible employment on different domains of psychosocial health. *J Epidemiol Commun Health* 59(9):761–767.

Bambra, C., T. Lunau, K. A. Van der Wel, T. A. Eikemo, and N. Dragano. 2014. Work, health and welfare: The association between working conditions, welfare states and self-reported general health in Europe. *Int J Health Serv* 44(1):113–136.

Bardasi, E., and M. Francesconi. 2004. The impact of atypical employment on individual well-being: Evidence from a panel of British workers. *Soc Sci Med* 58(9):1671–1688.

Benach, J., A. Vives, M. Amable, C. Vanroelen, G. Tarafa, and C. Muntaner. 2014. Precarious employment: Understanding an emerging social determinant of health. *Annu Rev Public Health* 35:229–253.

Bernhard-Oettel, C., M. Sverke, and H. De Witte. 2005. Comparing three alternative types of employment with permanent full-time work: How do employment contract and perceived job conditions relate to health complaints? *Work Stress* 19(4):301–318.

Bhandari, A. K., and A. Hesmati. 2006. Wage inequality and job insecurity among permanent and contract workers in India: Evidence from organized manufacturing industries. *IZA Discussion Paper No. 2097.* http://ftp.iza.org/dp2097.pdf. (accessed January 8, 2020).

Blau, P. 1964. *Exchange and power in social life*. New York, NY: Wiley.

Blomme, R. J., A. van Rheede, and D. M. Tromp. 2010. The use of the psychological contract to explain turnover intentions in the hospitality industry: A research study on the impact of gender on the turnover intentions of highly educated employees. *Int J Hum Resour Manag* 21(1):144–162.

Bordia, P., S. L. Restubog, and R. L. Tang. 2008. When employees strike back: Investigating the mediating mechanisms between psychological contract breach and workplace deviance. *J Appl Psychol* 93(5):1104–1117.

Bugajska, J., T. Makowiec-Dąbrowska, A. Bortkiewicz et al. 2011. Physical capacity of occupationally active population and possibilities to perform hard physical work. *Int J Occup Saf Ergon* 11(2):129–138.

Callea, A., F. Urbini, E. Ingusci, and A. Chirumbolo. 2016. The relationship between contract type and job satisfaction in a mediated moderation model: The role of job insecurity and psychological contract violation. *Econ Industr Democr* 37(2):399–420.

Canivet, C., T. Bodin, M. Emmelin, S. Toivanen, M. Moghaddassi, and P. O. Östergren. 2016. Precarious employment is a risk factor for poor mental health in young individuals in Sweden: A cohort study with multiple follow-ups. *BMC Public Health* 16:687. https://www.researchgate.net/publication/305801182_Precarious_employment_is_a_risk_factor_for_poor_mental_health_in_young_individuals_in_Sweden_A_cohort_study_with_multiple_follow-ups. (accessed January 8, 2020).

Cappelli, P., and J. Keller. 2013. Classifying work in the new economy. *Acad Manag Rev* 38(4):575–596.

Cassar, V., and R. Briner. 2011. The relationship between psychological contract breach and organizational commitment: Exchange imbalance as a moderator of the mediating role of violation. *J Vocat Behav* 78(2):283–289.

Chambel, M. J., and F. Castanheira. 2006. Different temporary work status: Different behaviors in organization. *J Bus Psychol* 20(3):351–367.

Chambel, M. J., and F. Sobral. 2011. Training is an investment with return in temporary workers: A social exchange perspective. *Career Develop Int* 16(2):161–177.

Chłoń-Domińczak, A., and M. Palczyńska. 2015. *Rynek pracy a kompetencje Polaków: Wybrane wyniki badania postPIAAC* [Labour market and Poles competencies: Selected results of the postPIAAC study]. Warszawa: Instytut Badań Edukacyjnych. http://produkty.ibe.edu.pl/docs/raporty/ibe-ee-raport-postpiaac.pdf. (accessed January 8, 2020).

Cichocki, S., K. Saczuk, P. Strzelecki, J. Tyrowicz, and R. Wyszyński. 2013. Kwartalny raport o rynku pracy: I kwartał 2013 r [Quarterly report on the labor market: Q1 2013]. Narodowy Bank Polski. Instytut Ekonomiczny. https://www.nbp.pl/publikacje/rynek_pracy/rynek_pracy_2013_1kw.pdf. (accessed January 8, 2020).

Conger, J. A., K. Kanuago, and S. T. Menon. 2000. Charismatic leadership and follower effects. *J Organ Behav* 21:747–767.

Connelly, C. E., and D. G. Gallagher. 2006. Independent and dependent contracting: Meaning and implications. *Hum Resour Manag Rev* 16(2):95–106.

Conway, N., and R. B. Briner. 2005. *Understanding psychological contracts at work: A critical evaluation of theory and research.* Oxford, UK: Oxford University Press.

Conway, N., and J. A. M. Coyle-Shapiro. 2012. The reciprocal relationship between psychological contract fulfilment and employee performance and the moderating role of perceived organizational support and tenure. *J Occup Organ Psychol* 85(2):277–299.

Coyle-Shapiro, J. A. M., and I. Kessler. 2000. Consequences of the psychological contract for the employment relationship. A large-scale study. *J Manag Stud* 37(7):903–930.

Coyle-Shapiro, J. A. M., and I. Kessler. 2002. Reciprocity through the lens of the psychological contract: Employee and employer perspectives. *Eur J Work Organiz Psychol* 11(1):1–18.

Dawson, C., M. Veliziotis, G. Pacheco, and D. J. Webber. 2005. Is temporary employment a cause or consequence of poor mental health? A panel data analysis. *Soc Sc Med* 134:50–58.

De Cuyper, N., J. de Jong, H. De Witte, K. Issakson, T. Rigotti, and R. Schalk. 2008b. Literature review of theory and research on the psychological impact of temporary employment: Towards a conceptual model. *Int J Manag Rev* 10(1):25–51.

De Cuyper, N., and H. De Witte. 2006a. The impact of job insecurity and contract type on attitudes, well-being and behavioural reports: A psychological contract perspective. *J Occup Organiz Psychol* 79(3):395–409.

De Cuyper, N., and H. De Witte. 2006b. Autonomy and workload among temporary workers: Their effects on job satisfaction, organizational commitment, life satisfaction and self-rated performance. *Int J Stress Manag* 13(4):441–459.

De Cuyper, N., and H. De Witte. 2007a. Job insecurity in temporary versus permanent workers: Associations with attitudes, well-being, and behaviour. *Work Stress* 21(1):65–84.

De Cuyper, N., and H. De Witte. 2007b. Associations between contract preference and attitudes, wellbeing and behavioural intentions of temporary workers. *Econ Industr Democr* 28(2):292–312.

De Cuyper, N., and H. De Witte. 2008a. Volition and reasons for accepting temporary employment: Associations with attitudes, well-being, and behavioural intentions. *Eur J Work Organiz Psychol* 17(3):363–387.

De Cuyper, N., H. De Witte, and H. Van Emmerik. 2011. Temporary employment: Costs and benefits for (the careers of) employees and organizations. *Car Develop Int* 16(2):104–113.

De Cuyper, N., T. Rigotti, H. De Witte, and G. Mohr. 2008c. Balancing psychological contracts: Validation of a typology. *Int J Hum Resour Manag* 19(4):543–561.

De Gilder, D. 2003. Commitment, trust and work behavior: The case of contingent workers. *Person Rev* 32(2):588–604.

De Jong, J., N. De Cuyper, H. De Witte, I. Silla, and C. Bernhard-Oettel. 2009. Motives for accepting temporary employment: A typology. *Int J Manpow* 30(3):237–252.

De Jong, J., and R. Schalk. 2005. Temporary employment in the Netherlands: Between flexibility and Security. In: *Employment contracts and well-being among European workers*, eds. N. De Cuyper, K. Isaksson, and H. De Witte, 119–152. Aldershot, UK: Ashgate.

De Jong, J., and R. Schalk. 2010. Extrinsic motives as moderators in the relationship between fairness and work-related outcomes among temporary workers. *J Bus Psychol* 25(1):175–189.

De Witte, H., and K. Näswall. 2003. Objective versus subjective job insecurity: Consequences of temporary work for job satisfaction and organizational commitment in four European countries. *Econ Industr Democr* 24(2):149–188.

Dineen, B., and D. Allen. 2016. Third party employment branding: Human capital inflows and outflows following "best places to work" certifications. *Acad Manag J* 59(1):90–112.

Dixon, J. C., A. S. Fullerton, and D. L. Robertson. 2013. Cross-national differences in workers' perceived job, labour market, and employment insecurity in Europe: Empirical tests and theoretical extensions. *Eur Soc Rev* 29(5):1053–1067.

Duchaine, C. S., R. Ndjaboué, M. Levesque et al. 2017. Psychosocial work factors and social inequalities in psychological distress: A population-based study. *BMC Public Health* 17(1):91. https://bmcpublichealth.biomedcentral.com/articles/10.1186/s12889-017-4014-4. (accessed January 8, 2020).

Ellingson, J. E., M. L. Gruys, and P. R. Sackett. 1998. Factors related to the satisfaction and performance of temporary employees. *J Appl Psychol* 83(6):913–921.

Engellandt, A., and R. T. Riphahn. 2005. Temporary contracts and employee efforts. *Lab Econ* 12(3):281–299.

Ervasti, J., J. Vahtera, P. Virtanen et al. 2014. Is temporary employment a risk factor for work disability due to depressive disorders and delayed return to work? The Finnish Public Sector Study. *Scand J Work Environ Health* 40(4):343–352.

Eurofound. 2015. *Recent developments in temporary employment: Employment growth, wages and transitions*. Luxembourg: Publications Office of the European Union. https://www.researchgate.net/publication/327155958_Recent_developments_in_temporary_employment_Employment_growth_wages_and_transitions. (accessed January 8, 2020).

Eurofound. 2017. *Aspects of non-standard employment in Europe*. Luxembourg: Publications Office of the European Union. https://www.eurofound.europa.eu/sites/default/files/ef_publication/field_ef_document/ef1724en.pdf. (accessed January 8, 2020).

Eurofound. 2018. *Non-standard forms of employment: Recent trends and future prospects*. Luxembourg: Publications Office of the European Union. https://www.eurofound.europa.eu/sites/default/files/ef_publication/field_ef_document/ef1746en.pdf.

Farrell, D., and F. Greig. 2016. *Paychecks, paydays, and the online platform economy: Big data on income volatility*. New York, NY: JPMorgan Chase & Co. Inst. https://www.jpmorganchase.com/corporate/institute/document/jpmc-institute-volatility-2-report.pdf. (accessed January 8, 2020).

Feldt, T., K. Hyvönen, A. Mäkikangas, U. Kinnunen, and K. Kokko. 2009. Development trajectories of Finnish manager's work ability over a 10-year period. *Scand J Work Environ Health* 35(1):37–47.

Forrier, A., and L. Sels. 2003. Temporary employment and employability: Training opportunities and efforts of temporary and permanent employees in Belgium. *Work Employ Soc* 17(4):641–666.

Fulmer, I., B. Gerhart, and K. Scott. 2003. Are the 100 best better? An empirical investigation of the relationship between being a "great place to work" and firm performance. *Pers Psychol* 56(4):965–993.

Gallagher, D. G., and J. McLean Parks. 2001. I pledge thee my troth contingently: Commitment and the contingent work relationship. *Hum Resour Manag Rev* 11(3):181–208.

Goldberg, D. P., and P. Williams. 1991. *A user guide to the general health questionnaire*. Berkshire: NFER-NELSON.

Goraus, K., and P. Lewandowski. 2016. Minimum wage violation in Central and Eastern Europe. *IBS Working Paper 3/2016*. https://ibs.org.pl/app/uploads/2016/04/IBS_Working_Paper_03_2016.pdf. (accessed January 8, 2020).

Goudswaard, A., and F. Andries. 2002. *Employment status and working conditions*. Luxembourg: European Foundation for the Improvement of Working and Living Conditions. Office for Official Publications of The European Community.

Gracia, F. J., J. Ramos, J. M. Peiro, A. Caballer, and B. Sora. 2011. Job attitudes, behaviours and well-being among different types of temporary workers in Europe and Israel. *Int Lab Rev* 150(3–4):235–254.

Guest, D. 2004. Flexible employment contracts, the psychological contract and employee outcomes: An analysis and review of the evidence. *Int J Manag Rev* 5/6(1):1–19.

Guest, D. 2017. Human resource management and employee well-being: Towards a new analytic framework. *Hum Resour Manag J* 27(1):22–38.

Guest, D., and M. Clinton. 2006. *Temporary employment contracts, workers' wellbeing and behavior: Evidence from the U. K.* London: Department of Management Working Paper, N. 38. https://www.researchgate.net/publication/233590524_Temporary_Employment_Associations_with_Employees'_Attitudes_Well-Being_and_Behaviour_A_Review:of_Belgian_Research. (accessed January 8, 2020).

Guest, D., K. Isaksson, and H. de Witte. 2010. *Employment contracts, psychological contracts and employee well-being: An international study*. Oxford: Oxford University Press.

Hammarström, A., P. Virtanen, and U. Janlert. 2011. Are the health consequences of temporary employment worse among low educated than among high educated? *Eur J Pub Health* 21(6):756–776.

Han, K. M., J. Chang, E. Won, M. S. Lee, and B. J. Ham. 2017. Precarious employment associated with depressive symptoms and suicidal ideation in adult wage workers. *J Affect Disord* 218:201–209.

Hasselhorn, H. M. 2008. *Work ability: Concept and assessment.* http://www.arbeitsfaehigkeit.uni-wuppertal.de/picture/upload/file/Concept_and_Assessment.pdf. (accessed January 9, 2020).

Hekman, D. R., G. A. Bigley, K. Steensma, and J. F. Hereford. 2009. Combined effects of organizational and professional identification on the reciprocity dynamic for professional employees. *Acad Manag J* 52(3):506–526.

Hu, Y., F. J. van Lenthe, G. J. Borsboom et al. 2016. Trends in socioeconomic inequalities in self-assessed health in 17 European countries between 1990 and 2010. *J Epidemiol Commun Health* 70(7):644–652.

Ilmarinen, J., and K. Tuomi. 2004. *Past, present and future of workability: People and work research reports*, vol. 65, pp. 1–25. Helsinki: Finnish Institute of Occupational Health.

ILO (International Labour Organization). 2016. *Non-standard employment around the world: Understanding challenges, shaping prospects.* Geneva: International Labour Office. https://www.ilo.org/wcmsp5/groups/public/---dgreports/---dcomm/---publ/documents/publication/wcms_534326.pdf. (accessed January 8, 2020).

Imhof, S., and M. Andresen. 2018. Unhappy with well-being research in the temporary work context: Mapping review and research agenda. *Int J Hum Resour Manag* 29(1):127–164.

Isaksson, K., and K. Bellagh. 2002. Health problems and quitting among female "temps". *Eur J Work Organiz Psychol* 11(1):27–45.

Isaksson, K., N. De Cuyper, C. Bernhard Oettel, and H. De Witte. 2010. The role of the formal employment contract in the range and fulfilment of the psychological contract: Testing a layered model. *Eur J Work Organiz Psychol* 19(6):696–716.

Jang, S. Y., S. I. Jang, H. C. Bae, J. Shin, and E. Park. 2015. Precarious employment and new-onset severe depressive symptoms: A population-based prospective study in South Korea. *Scand J Work Environ Health* 41(4):329–337.

Kachi, Y., T. Otsuka, and T. Kawada. 2014. Precarious employment and the risk of serious psychological distress: A population-based cohort study in Japan. *Scand J Work Environ Health* 40(5):465–472.

Kalleberg, A. L. 2009. Precarious work, insecure workers: Employment relations in transition. *Americ Soc Rev* 74(1):1–22.

Kalleberg, A. L. 2001. Organizing flexibility: The flexible firm in a new century. *British J Industr Rel* 39(41):479–504.

Kalleberg, A. L. 2011. *Good jobs, bad jobs: The rise of polarized and precarious employment systems in the United States, 1970s–2000s.* New York, NY: Russell Sage Foundation.

Kammeyer-Mueller, J., and H. Liao. 2006. Workforce reduction and job-seeker attraction: Examining job seekers' reactions to firm workforce-reduction policies. *Hum Resour Manage* 45(4):585–603.

Katz, L. F., and A. B. Krueger. 2016. *The rise and nature of alternative work arrangements in the United States, 1995–2015.* Cambridge, MA: National Bureau of Economic Research. https://www.nber.org/papers/w22667.pdf. (accessed January 8, 2020).

Kauhanen, M., and J. Nätti. 2015. Involuntary temporary and part-time work, job quality and well-being at work. *Soc Indic Res* 120(3):783–799.

Kiewitz, C., S. L. D. Restubog, T. Zagenczyk, and W. Hochwarter. 2009. The interactive effects of psychological contract breach and organizational politics on perceived organizational support: Evidence from two longitudinal studies. *J Manag Stud* 46(5):806–834.

Kim, M. H., C. Y. Kim, J. K. Park, and I. Kawachi. 2008. Is precarious employment damaging to self-rated health? Results of propensity score matching methods, using longitudinal data in South Korea. *Soc Sci Med* 67(12):1982–1994.

Kim, I. H., C. Muntaner, F. Vahid Shahidi, A. Vives, C. Vanroelen, and J. Benach. 2012. Welfare states, flexible employment, and health: A critical review. *Health Pol* 104(2):99–127.

Kim, W., E. C. Park, T. H. Lee, and T. H. Kim. 2016. Effect of working hours and precarious employment on depressive symptoms in South Korean employees: A longitudinal study. *Occup Environ Med (OEM)* 73(12):816–822.

Kinnunen, U., A. Mäkikangas, S. Mauno, K. Siponen, K. J. Järvinen, and J. Nätti. 2011. Perceived employability: Investigating outcomes among involuntary and voluntary temporary employees compared to permanent employees. *Car Develop Int* 16(2):140–160.

Kline, R. B. 2005. *Principles and practice of structural equation modeling*. 2nd ed. New York, NY: The Guilford Press.

Krausz, M., A. Sagie, and Y. Bidermann. 2000. Actual and preferred work schedules and scheduling control as determinants of job-related attitudes. *J Vocat Behav* 56(1):1–11.

Kuvaas, B. 2008. An exploration of how the employee-organization relationship affects the linkage between perception of developmental human resource practices and employee outcomes. *J Manag Stud* 45:1–25.

Lahelma, E., M. Laaksonen, T. Lallukka et al. 2012. Working conditions as risk factors for disability retirement: A longitudinal register linkage study. *BMC Public Health* 12(1):309. https://bmcpublichealth.biomedcentral.com/articles/10.1186/1471-2458-12-309. (accessed January 8, 2020).

Lambert, L. S. 2011. Promised and delivered inducements and contributions: An integrative view of psychological contract appraisal. *J Appl Psychol* 96(4):695–712.

Leinonen, T., O. Pietiläinen, M. Laaksonen, O. Rahkonen, E. Lahelma, and P. Martikainen. 2011. Occupational social class and disability retirement among municipal employees: The contribution of health behaviors and working conditions. *Scand J Work Environ Health* 37(6):464–472.

Lewandowski, P., M. Góra, and M. Lis. 2017. *Temporary employment boom in Poland: A job quality vs quantity trade-off.* Bonn: Institute of Labor Economics. https://ibs.org.pl/app/uploads/2017/09/IBS_Working_Paper_04_2017.pdf. (accessed January 8, 2020).

Lopes, S., and M. J. Chambel. 2017. Temporary agency workers' motivations and well-being at work: A two-wave study. *Int J Stress Manag* 24(4):321–346.

Lunde, L. K., M. Koch, S. Knardahl et al. 2014. Musculoskeletal health and work ability in physically demanding occupations: Study protocol for a prospective field study on construction and health care workers. *BMC Public Health* 14:1075.

Makowska, Z., and D. Merecz. 2001. *Ocena zdrowia psychicznego na podstawie badań kwestionariuszami Dawida Goldberga* [Mental Health Evaluation Using David Goldberg Tool]. Łódź: Instytut Medycyny Pracy.

Marler, J. H., M. W. Barringer, and G. T. Milkovich. 2002. Boundaryless and traditional contingent employees: Worlds apart. *J Organ Behav* 23:425–453.

Martens, M. F. J., F. J. N. Nijhuis, M. P. J. van Boxtel, and J. A. Knottnerus. 1999. Flexible work schedules and mental and physical health: A study of a working population with non-traditional working hours. *J Organ Behav* 20(1):35–46.

Mc Donald, D. J., and P. J. Makin. 2000. The psychological contract, organizational commitment and job satisfaction of temporary staff. *Lead Organiz Develop J* 21:84–91.

McLean Parks, J., D. L. Kidder, and D. G. Gallagher. 1998. Fitting square pegs into round holes: Mapping the domain of contingent work arrangements onto the psychological contract. *J Organiz Behav* 19:697–730.

Meyer, A., and M. Wallette. 2005. *Absence of absenteeism and overtime work: Signalling factors of temporary workers?* Lund: Lund University. Department of Economics. https://project.nek.lu.se/publications/workpap/Papers/WP05_15.pdf. (accessed January 8, 2020).

Moscone, F., E. Tosetti, and G. Vittadini. 2016. The impact of precarious employment on mental health: The case of Italy. *Soc Sci Med* 158:86–95.

Niedhammer, I., T. Lesuffleur, T. Coutrot, and J. F. Chastang. 2016. Contribution of working conditions to occupational inequalities in depressive symptoms, results from the national French SUMER survey. *Int Arch Occup Environ Health* 89(6):1025–1037. DOI: 10.1007/s00420-016-1142-6.pdf.

Nolen, S. D. 1996. Negative aspects of temporary employment. *J Lab Res* 17(4):567–582.

Nunez, I., and I. Livanos. 2011. Temps "by choice"? an analysis of reasons for temporary employment among young European workers. *Paper presented at ESRC seminar "Young workers and precarious employment" September 23rd University of Warwick.* https://www.academia.edu/1138835/Temps_by_choice_An_Analysis_of_reasons_for_temporary_employment_among_young_European_workers. (accessed January 8, 2020).

OECD [Organisation for Economic Co-operation and Development]. 2019. *Employment outlook the future of work.* Paris: OECD Publishing. DOI: 10.1787/9789264306943-en.

Oliv, S., N. E. Gustafsson, and M. A. Hagberg. 2017. A Lower level of physically demanding work is associated with excellent work ability in men and women with neck pain in different age groups. *Saf Health Work* 8(4):356–363.

Pańków, M. 2015. Polityka państwa wobec upowszechnienia elastycznych form zatrudnienia w Polsce [State policy towards flexible employment implementation]. *Stud Pol Publicznej* 3(7):140–158.

Pirani, E. 2017. On the relationship between atypical work(s) and mental health: New insights from the Italian case. *Soc Indicat Res* 130(1):233–252.

Pirani, E., and S. Salvini. 2015. Is temporary employment damaging to health? A longitudinal study on Italian workers. *Soc Sci Med* 124:121–131.

Preacher, K. J., P. J. Curran, and D. J. Bauer. 2006. Computational tools for probing interaction effects in multiple linear regression, multilevel modeling, and latent curve analysis. *J Educ Behav Stat* 31(4):437–448.

Preacher, K. J., and A. F. Hayes. 2008. Asymptotic and resampling strategies for assessing and comparing indirect effects in multiple mediator models. *Behav Res Method* 40(3):879–891.

Quesnel-Vallée, A., S. DeHaney, and A. Ciampi. 2010. Temporary work and depressive symptoms: A propensity score analysis. *Soc Sci Med* 70(12):1982–1987. DOI: 10.1016/j.socscimed.2010.02.008.

Raja, U., G. Johns, and F. Ntalianis. 2004. The impact of personality on psychological contracts. *Acad Manag J* 47:350–367.

Ray, T. K., T. A. Kenigsberg, and R. Pana-Cryan. 2017. Employment arrangement, job stress, and health-related quality of life. *Saf Sci* 100(A):46–56.

Rigotti, T., and G. Mohr. 2005. German flexibility: Loosening the reins without losing control. In: *Employment contracts and well-being among European workers*, eds. N. De Cuyper, K. Isaksson, and H. De Witte, 75–102. Aldershot, UK: Ashgate.

Robinson, S. L., and D. M. Rousseau. 1994. Violating the psychological contract: Not the exception but the norm. *J Organiz Behav* 15(3):245–259. DOI: 10.1002/job.4030150306.

Rodriguez, E. 2002. Marginal employment and health in Britain and Germany: Does unstable employment predict health? *Soc Sci Med* 55(6):963–979.

Roelen, C. A. M., M. F. A. van Hoffen, S. Waage et al. 2018. Psychosocial work environment and mental health-related long-term sickness absence among nurses. *Int Arch Occup Environ Health* 91(2):195–203.

Rousseau, D. M. 1990. New hire perceptions of their own and their employer's obligations: A study of psychological contracts. *J Organiz Behav* 11(5):389–400. DOI: 10.1002/job.4030110506.

Rousseau, D. M. 1995. *Psychological contracts in organizations: Understanding written and unwritten agreements.* Newbury Park, CA: Sage.

Rousseau, D. M. 2004. Psychological contracts in the workplace: Understanding the ties that motivate. *Ac Manag Exec* 18(1):120–127. DOI: 10.5465/AME.2004.12689213.

Rousseau, D. M., S. D. Hansen, and M. Tomprou. 2018. A dynamic phase model of psychological contract processes. *J Organiz Behav.* DOI: 10.1002/job.2284.

Rousseau, D. M., and S. A. Tijoriwala. 1998. Assessing psychological contracts: Issues, alternatives, and types of measures. *J Organ Behav* 19:679–695.

Schaufeli, W., and A. Baker. 2003. *Utrecht work engagement scale: Preliminary manual.* Utrecht: Occupational Health Psychology Unit Utrecht University. https://www.wilmarschaufeli.nl/publications/Schaufeli/Test%20Manuals/Test_manual_UWES_English.pdf. (accessed January 8, 2020).

Schein, E. 1978. *Career dynamics: Matching individual and organizational needs.* Reading, MA: Addison-Wesley.

Schouten, L. S., C. I. Joling, J. W. J. van der Gulden, M. W. Heymans, U. Bultmann, and C. A. M. Roelen. 2015. Screening manual and office workers for risk of long-term sickness absence: Cut-off points for the Work Ability Index. *Scand J Work Environ Health* 41(1):36–42. https://www.sjweh.fi/show:abstract.php?abstract_id=3465&fullText=1#box-fullText (accessed January 8, 2020).

Schütte, S., J. F. Chastang, A. Parent-Thirion, G. Vermeylen, and I. Niedhammer. 2015. Psychosocial work exposures among European employees: Explanations for occupational inequalities in mental health. *J Public Health (Oxf)* 37(3):373–388.

Snape, E., and T. Redman. 2010. HRM practices, organizational citizenship behaviour, and performance: A multi-level analysis. *J Manag Stud* 47:1219–1247.

Standing, G. 2011. *The precariat: The new dangerous class.* London: Bloomsbury Academic.

Sverke, M., J. Hellgren, and K. Näswall. 2002. No security: A meta-analysis and review of job insecurity and its consequences. *J Occup Health Psychol* 7(3):242–264.

Szabowska-Walaszczyk, A., A. M. Zalewska, and M. Wojtaś. 2011. Zaangażowanie w pracę i jego korelaty: Adaptacja skali UWES autorstwa Schaufeliego i Bakkera [Work engagement and its correlates: The adaptation of the UWES by Schauffeli and Baker scale]. *Psychol Jakości życia* 10:57–74.

Tan, H. H., and C. P. Tan. 2002. Temporary employees in Singapore: What drives them? *J Psychol Interd Appl* 136(1):83–102.

Tekleab, A. G., and S. Taylor. 2003. Aren't two parties in an employment relationship? Antecedents and consequences of organization-employee agreement on contract obligations and violations. *J Organiz Behav* 24(5):585–608.

Thorsen, S. V., H. Burr, F. Diderichsen, and J. B. Bjorner. 2013. A one-item workability measure mediates work demands, individual resources and health in the prediction of sickness absence. *Int Arch Occup Environ Health* 86(7):755–766.

Tonnon, S. C., S. R. J. Robroek, A. J. van der Beek, A. Burdorf, H. P. van der Ploeg, M. Caspers, and K. I. Proper. 2019. Physical workload and obesity have a synergistic effect on work ability among construction workers. *Int Arch Occup Environ Health* 92(6):855–864.

Tuomi, K., L. Eskelinen, J. Toikkanen, E. Jarvinen, J. Ilmarinen, and M. Klockars. 1991. Work load and individual factors affecting work ability among aging municipal employees. *Scand J Work Environ Health* 17(1):128–134.

Turnley, W. H., and D. C. Feldman. 1999. The impact of psychological contract violation on exit, voice, loyalty, and neglect. *Hum Relat* 52(7):895–922.

Uen, J. F., M. S. Chien, and Y. F. Yen. 2009. The mediating effects of psychological contracts on the relationship between human resource systems and role behaviors: A multi-level analysis. *J Bus Psychol* 24(2):215–223.

Vahtera, J., P. Virtanen, M. Kivimäki, and J. Pentti. 1999. Workplace as an origin of health inequalities. *J Epidemiol Commun Health* 53(7):399–407.

Van Aerden, K., G. Moors, K. Levecque, and C. Vanroelen. 2015. The relationship between employment quality and work-related well-being in the European Labor Force. *J Vocat Behav* 86:66–76. DOI: 10.1016/j.jvb.2014.11.001.

Van Aerden, K., V. Puig-Barrachina, K. Bosmans, and C. Vanroelen. 2016. How does employment quality relate to health and job satisfaction in Europe? A typological approach. *Soc Sci Med* 158:132–140.

Van Breugel, G., W. Van Olffen, and R. Olie. 2005. Temporary liaisons: The commitment of temps towards their agencies. *J Manag Stud* 42(3):539–566.

Van Breukelen, W., and J. Allegro. 2000. Effecten van een nieuwe vorm van flexibilisering van de arbeid [Effects of a new form of labour flexibility]. *Gedrag Organ* 13(2):107–124. After De Cuyper, N., K. Isaksson, and H. De Witte. 2005. *Employment contracts and psychological contracts among European workers*. Aldershot, UK: Ashgate.

Van den Berg, T. I., L. A. Elders, B. C. de Zwart, and A. Burdorf. 2009. The effects of work-related and individual factors on the Work Ability Index: A systematic review. *Occup Environ Med* 66(4):211–220.

Virtanen, M., M. Kivimäki, M. Elovainio, and J. Vahtera. 2002. Selection from fixed-term to permanent employment: Prospective study on health, job satisfaction, and behavioural risks. *J Epidemiol Commun Health* 56(9):693–699.

Virtanen, M., M. Kivimäki, M. Elovainio, J. Vahtera, and C. L. Cooper. 2001. Contingent employment, health and sickness absence. *Scand J Work Environ Health* 27(6):365–372.

Virtanen, M., M. Kivimäki, M. Elovainio, J. Vahtera, and J. E. Ferrie. 2003a. From insecure to secure employment: Changes in work, health and health-related behaviours. *Occup Environ Med* 60(12):948–953.

Virtanen, M., M. Kivimäki, M. Joensuu, P. Virtanen, M. Elovainio, and J. Vahtera. 2005. Temporary employment and health: A review. *Int J Epidemiol* 34(3):610–622.

Virtanen, P., V. Liukkonen, J. Vahtera, M. Kivimäki, and M. Koskenvuo. 2003b. Health inequalities in the workforce: The labour market core-periphery structure. *Int J Epidemiol* 32(6):1015–1021.

Vives, A., M. Amable, M. Ferrer et al. 2013. Employment precariousness and poor mental health: Evidence from Spain on a new social determinant of health. *J Environ Public Health*:1–10. https://new.hindawi.com/journals/jeph/2013/978656/#. (accessed January 8, 2020).

Vives, J. M. A., G. Tarafa, and J. Benach. 2017. Changing the way we understand precarious employment and health: Precarisation affects the entire salaried population. *Saf Sci* 100:66–73.

Wood, A. J. 2016. Flexible scheduling, degradation of job quality and barriers to collective voice. *Hum Relat* 69(10):1989–2010.

Wooden, M. 2004. Non-standard employment and job satisfaction: Evidence from the HILDA survey. *J Ind Relat* 46(3):275–297.

Zeytinoglu, I. U., and G. B. Cooke. 2005. Non-standard work and benefits: Has anything changed since the Wallace report? *Ind Relat (Berkeley)* 60:29–63.

Zhao, H., S. J. Wayne, B. Glibkowski, and J. Bravo. 2007. The impact of psychological contract breach on work-related outcomes: A meta-analysis. *Pers Psychol* 60(3):647–680.

5 Why Are Employees Counterproductive? *The Role of Social Stressors, Job Burnout and Job Resources*

Łukasz Baka

CONTENTS

5.1 INTRODUCTION

The research on occupational stress conducted over the past 60 years has shown that the negative effects of work-related stress are particularly pronounced in two areas: mental health and the organizational behaviors of employees. Regarding the former aspect, numerous empirical studies have unanimously proved that working environment stressors are a source of diverse somatic ailments (e.g. cardiovascular problems, muscle and bone pains) and mental health problems (e.g. burnout and depression).

The negative impact of work-related stressors may be mitigated by both the characteristics of the working environment (e.g. social support by management and co-workers), and the individual characteristics of employees (e.g. temperament type). In turn, the second area concerning the relationship between work-related stressors and negative organizational behaviors seems to have been less studied. Over the past decades, *counterproductive work behavior (CWB)*, defined as voluntary actions that harm or are intended to harm an organization or people associated with it – e.g. management, co-workers, clients – has been widely explored, in particular by American researchers in the field.

According to the *stressor-emotion model* (Spector and Fox 2005), the primary sources of CWB are stressors occurring in the working environment, notably those related to social relations, e.g. interpersonal conflicts and workplace bullying (Bruk-Lee and Spector 2006; Kessler et al. 2013), and conducive to strong negative emotions, e.g. anger or hostility. Counterproductive work behavior is a form of releasing these emotions and taking revenge on the organization for the "bad" treatment. The mediating role of negative emotions has been widely supported by empirical evidence (e.g. Bauer and Spector 2015; Fox et al. 2001; Penney and Spector 2007). It seems, however, that apart from negative emotions at work, which are a direct and relatively short-term reaction to stressful situations, there are other, somewhat more stable over time, mediating factors that the stress-emotion model does not account for. An example of such a factor is job burnout. Previous research has shown that burnout develops as a result of a long-term, chronic stress at work (e.g. Maslach et al. 2001) and is positively associated with various types of counterproductive work behaviors (e.g. Banks et al. 2012; Luksyte et al. 2011). In addition, the impact of social stressors may be weakened by individual job resources – e.g. job control and social support at work (Fox et al. 2001). The aim of this chapter is to examine (1) the direct relationship between social stressors and CWB; (2) the mediational effect of job burnout and (3) the moderating effect of job resources on the social stressors and CWB link. Two types of social stressors have been considered, i.e. interpersonal conflicts at work and workplace bullying, and two types of job resources, i.e. social support and job control.

5.1.1 Counterproductive Work Behaviors

A diverse taxonomy has been used in the literature to describe harmful behaviors at work. For example, these have been labeled *organizational aggression* (Neuman and Baron 1998), *antisocial behavior* (Giacalone and Greenberg 1997), *criminal behavior* (Hogan and Hogan 1989), *deviance work behavior* (Robinson and Bennett 1995) and *organizational retaliatory behavior* (Skarlicki and Folger 1997) or *revenge behavior* (Bies and Tripp 2005). Differences in terminology reflect the distinct theoretical approaches of the authors. For example, Neuman and Baron (Neuman 1998) derived their term from social psychology literature on aggression. Hogan and Hogan (1989) drew inspiration from criminology literature. Robinson and Bennett (Robinson and Bennett 1995) emphasized the violation of norms and organizational principles, whereas Skarlicki and Folger (1997) referred to the theory of justice. Despite these differences, the abovementioned behaviors are treated as examples of activities pertaining to a broadly defined category of negative organizational behaviors, hereinafter referred to as counterproductive work behavior. The latter term constitutes a *terminological umbrella* construct (Spector and Fox 2005), covering the aforementioned behaviors.

5.1.2 Typology of Counterproductive Work Behavior

Richard Hollinger and John Clark (Hollinger 1983) seem to have been pioneers of the first typology of counterproductive work behavior. Following an extensive study conducted on a large sample of employees based at three industrial plants, the researchers distinguished two general categories – *property deviance* and *production deviance*. The first group included thefts, destruction of property and abuse of privileges. The second category included various forms of non-compliance with work schedules (delays, extended breaks, leaving the workplace, leaving earlier), and activities reducing productivity (e.g. deliberate work delays, sluggishness, drinking alcohol at work). Both categories therefore corresponded to activities that hindered the achievement of organizational goals.

A decade later, two American researchers, Sandra Robinson and Rebecca Bennett, further developed the typology proposed by Hollinger and Clark (Hollinger 1983) by introducing an additional category of negative interpersonal behavior toward managers and co-workers (Bennett and Robinson 2000; Robinson and Bennett 1995). Based on advanced statistical methods, the authors distinguished two dimensions of CWB (Figure 5.1). The first dimension refers to the direction of counterproductive work behaviors. At the top end of the spectrum there are counterproductive work behaviors aimed at the organization as a whole, whereas at the bottom end there are behaviors directed at people connected with the organization. The second dimension concerns the degree of harm caused by CWB, whereby there have been identified minor and more serious counterproductive work behaviors located on opposite sides. The result is a matrix covering four CWB groups: production deviance, property deviance, political deviance, and personal aggression. In each group, the authors identified four types of counterproductive work behaviors, which yet make up only basic examples of such behaviors, and the full list is more exhaustive.

A more recent CWB classification has been proposed by researchers at the University of Florida under the supervision of Paul Spector and Suzy Fox (Spector 2005). Following a literature review and the authors' unique study results, the researchers distinguished five categories of counterproductive work behaviors – *abuse*, *sabotage*, *production deviance*, *theft* and *organizational withdrawal*. Table 5.1 presents a brief overview of the five types of counterproductive work behaviors.

Spector et al. (2006) introduced a distinction between active and passive CWB. The active forms include abuse, theft and sabotage, whereas the passive CWBs comprise production deviance and organizational withdrawal. The main source of active CWB forms is the intention to retaliate against the organization for "harm" and "humiliation", or unfair treatment suffered in the organization. These behaviors are based on affective motives – they are driven by strong emotions of anger, hostility and frustration. Their purpose is to release tension or punish the organization. Active CWBs are usually targeted directly at the source of stress and negative emotions (e.g.

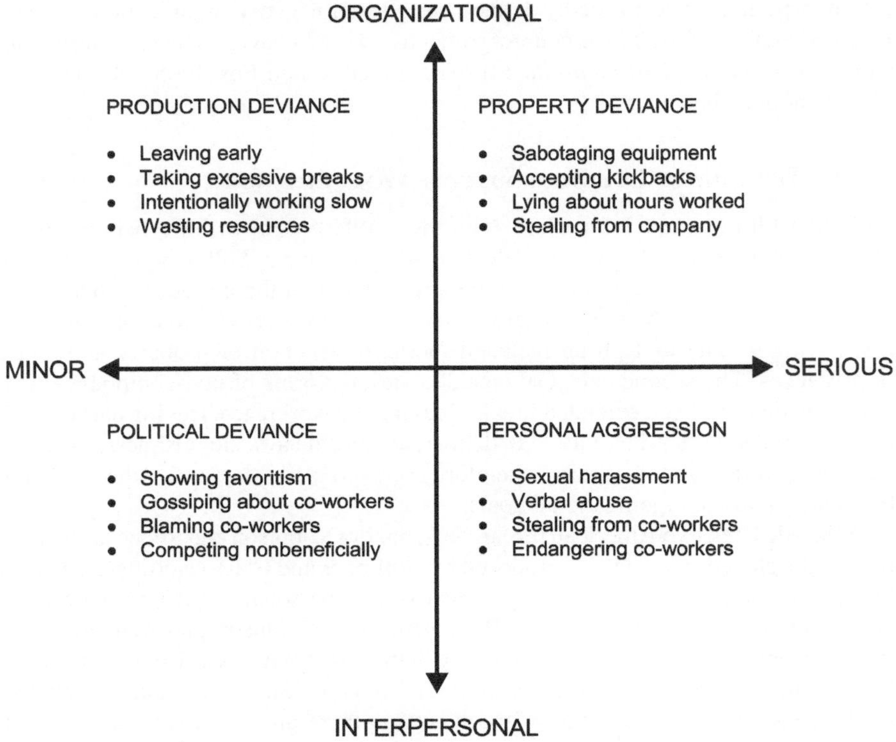

FIGURE 5.1 Typology of counterproductive work behavior by Robinson and Bennett (Robinson 1995).

managers, co-workers, organization's property). However, such behavior can be met with hostile reactions from managers and colleagues, as well as major organizational sanctions. This is notably the case when there is a significant disproportion in power relations between the manager and the employee. Therefore, in some cases, there are passive CWB variations – i.e. production deviance and organizational withdrawal.

5.1.3 Social Stressors as a Source of Counterproductive Work Behavior

At the end of the 1990s, two theoretical models were developed that sought to identify the mechanisms of counterproductive work behaviors, and built upon the premises of the theory of social exchange. These models are the *retaliatory behavior model* developed by the Canadian researchers Daniel Skarlicki and Robert Folger (Skarlicki 1997; Folger and Skarlicki 2005), and the *revenge model* proposed by the American researchers Robert Bies and Thomas Tripp (Bies 1997, 1998, 2005). Essentially, the two models are comparable to each other. Counterproductive work behavior is equally considered a behavioral response to "bad" interpersonal relations – e.g. unfair treatment by managers, psychological contract breach, interpersonal conflicts and experiences of workplace bullying or aggression at work. As the main

TABLE 5.1
Typology of Counterproductive Work Behaviors by Spector et al. (2006)

Abuse	Abuse of others is a violent behavior that aims to inflict physical or psychological harm on people associated with the organization. There are five types of work-related abuses – physical aggression (e.g. hitting, pushing), verbal aggression (e.g. shouting, calling, intimidating), offensive behaviors (e.g. offensive gestures, aggressive looks), ostracism (e.g. isolating someone, omitting someone's contribution, avoiding contact), instigation (e.g. persuading someone to carry out dangerous or forbidden activities). Abuse can also take the form of actively taking harmful actions or knowingly refraining from helpful actions or having a passive attitude.
Theft	Thefts are treated as a manifestation of aggressive behavior toward the organization as a whole. They consist of an unauthorized appropriation of company property by employees, or for their own use or sale to third parties. Some researchers also include time theft and deterioration in the quality of work in this definition. Thefts may be emotionally motivated and motivated by a desire to retaliate against the organization, or they may be driven by instrumental motives – the desire to obtain concrete benefits.
Sabotage	Sabotage is defined as deliberately harming, disrupting or boycotting the activities of the organization in order to achieve one's own goals. Sabotage involves both mild forms of behavior such as ignoring supervisor orders, deliberately delaying work, deliberately reducing the quality of work, dirtying and littering the workplace and creating a negative image of the company, as well as more severe forms of behavior such as deliberately destroying employer property, damaging equipment, violating laws and regulations, ignoring plans, misusing equipment or objects and using more materials than necessary.
Production deviance	Production deviance consists of an intentional decline of performance and quality of work, performing work in an inefficient way, failing to follow the recommendations and procedures and consciously making mistakes, as well as failing to report problems and abuses at the workplace to supervisors. While sabotage is treated as an active form of counterproductive work behavior, production deviance accounts for more passive forms. Because it is not targeted at specific employees, it is less visible on a daily basis, and thus more difficult to prove than sabotage. Sabotage and production deviance can be both emotional (e.g. a release of anger) and instrumental (e.g. forcing organizational changes).
Organizational withdrawal	Organizational withdrawal is a conscious undertaking of behaviors aimed at limiting the time spent on performing professional duties, as well as reducing the amount of energy spent on work. Such behaviors include intentional delay, shortening of working hours, extending breaks, leaving the workplace earlier, intentionally slowing down performance, taking days off that are not due and simulating illness. One of the manifestations of the withdrawal has also been increasingly described in literature, the phenomenon of *cyber loafing*, which involves surfing the Internet during working hours. This type of counterproductive work behavior consists of bringing losses to the organization through intentionally "doing nothing".

mechanism of these behaviors, the authors of both models identify the desire to take revenge for the harm suffered and punish the "unjust" organization.

These approaches refer to Adams' equilibrium (1965), a classical concept in psychology. It implies that employees more or less consciously compare their own balance of contributions and profits with that of contributions and profits of other employees. If this comparison turns out to be unfavorable, a subjective state of imbalance appears. The greater the imbalance, the greater the discomfort the employee feels, and the stronger the motivation to restore the balance. One way to restore balance is to engage in active counterproductive work behaviors, such as reducing effort, committing theft and damage to property or leaving the workplace unauthorized. This has been supported by both correlation and experimental studies. For example, Skarlicki and Folger (Skarlicki 1997) proved that an employee's response to unfair treatment in the organization can be a strong negative emotion (e.g. anger, rage, frustration), and a tendency to compensate for damages, e.g. thefts, acts of vandalism, organizational sabotage and reduction of work effort. In turn, Greenberg (1990) has shown that causing employees to feel injustice leads to increased theft and fluctuation in organization.

Previous research has also shown that over 90% of employees who have suffered harmful treatment from other co-workers or managers consider similar behavior as an available way of reciprocal action (Tripp and Bies 1997), and only 10% of employees cannot specify a negative personal reaction in their previous professional careers that would have been motivated by a desire to retaliate against the organization (Bies and Tripp 2005). Other studies have shown that revenge is socially acceptable to some extent. When the negative behavior is a reaction to a provocation or misconduct by other employees, it is deemed justified, even if it takes a drastic form and its consequences are more severe than those of the provocative behavior (Tripp et al. 2002). Particularly burdensome social stressors include interpersonal conflicts at work and bullying on account of managers.

5.1.3.1 Interpersonal Conflicts and Counterproductive Work Behavior

Interpersonal conflicts at work are an example of social stressors relating to the quality of relations among employees. They are based on negative, strenuous interactions between managers and co-workers. The friction often varies in intensity, from minor quarrels to a mental struggle (Spector and Jex 1998). Interpersonal conflicts can take many forms: open (e.g. open criticism or discrediting of an employee), or hidden (e.g. gossiping about co-workers), active (e.g. arguing, offensive comments), or passive (e.g. neglect, intentionally not answering phone calls).

In a survey study conducted by Keenan and Newton (Keenan 1985) engineers were asked to specify particularly stressful incidents that had taken place at the company. Among the identified situations, 74% related to social relations among managers, employees and co-workers, while interpersonal conflicts were the second most frequently mentioned stressor at work. Intercultural research has yielded similar results. In one such study, U.S. and Indian sales staff rated 11 stressors at work in terms of mental stress. Interpersonal conflict was the third highest rated stressor in the U.S. group and the fourth highest in the Indian group (Narayanan et al. 1999). Also, in a study conducted on an American–Chinese academic employee sample, interpersonal conflict was perceived as a significant stress factor (Liu et al. 2010). Interestingly, the U.S. employees were more likely to have conflicts with lower-level staff than with fellow colleagues, and these were conflicts of an open character. In turn, hidden interpersonal conflicts between co-workers were predominant among the Chinese employee group. The researchers observed differences in the consequences of interpersonal conflicts, depending on the nature of the conflict, such as a conflict between co-workers, or a difficult employee–manager relationship. While conflicts with co-workers led to more personal consequences (e.g. lowered mood, or self-esteem), conflicts with managers engendered organizational outcomes, such as diminished job satisfaction, or exacerbation of harmful behaviors at work. A strong relationship between interpersonal conflicts and counterproductive work behaviors has been demonstrated in several studies conducted in the United States, Turkey, Italy and Poland. Apart from several single studies, at least two study meta-analyses have confirmed the associations between the discussed phenomena. The corrected correlation coefficients of interpersonal conflicts and counterproductive work behaviors were $\rho = 0.38$; $p < 0.001$ (Hershcovis et al. 2007) and $\rho = 0.48$; $p < 0.001$ (Berry et al. 2012), respectively.

5.1.3.2 Workplace Bullying and Counterproductive Work Behavior

The literature in the field offers many accounts of research examining the relationship between the experience of workplace bullying and counterproductive work behaviors (Einarsen et al. 2003; Warszewska 2013). In a study conducted by Italian researchers (Balducci et al. 2011) the results revealed that workers experiencing workplace bullying were more likely to display harassing behaviors in relation to fellow co-workers ($r = 0.39$; $p < 0.001$). Bibi and Karim (Bibi 2013) have shown that workplace bullying is positively associated with the five dimensions of counterproductive work behavior identified by Spector and Fox (2005). The strongest correlation has been found between workplace bullying and organizational withdrawal ($r = 0.55$; $p < 0.001$), while the weakest was between workplace bullying and sabotage ($r = 0.25$; $p < 0.001$). A positive correlation between workplace bullying and general counterproductive work behaviors ($r = 0.26$) has also been observed in longitudinal studies with a two-month measurement interval (Sakurai and Jex 2012).

One of the forms of workplace bullying is *abusive supervision*, defined as employee subjectively perceived level of verbal and non-verbal hostility of the manager (Tepper 2000). Such management practice includes using offensive nicknames in relation to employees, yelling at employees, aggressive gestures, intimidation, mocking of mistakes, failing to share information, inciting dangerous or punishable activities and isolation. Numerous studies have proven that degrading treatment on behalf of management results in frustration and intensification of retaliatory tendencies among employees (e.g. Mitchell and Ambrose 2007). For example, in a study conducted by Mitchell and Ambrose (2007), abusive supervision behaviors were associated with an increase in counterproductive work behavior directed both directly toward the manager ($r = 0.40$), other co-workers ($r = 0.17$) and at the organization as a whole ($r = 0.20$).

Over the last few years, several longitudinal studies have been carried out in which researchers have examined the relationship between abusive supervision and counterproductive employee behavior (Simon 2015; Wei and Si 2013). In one such study, six measurements were conducted over six months, whereby the experience of abuse of authority by a manager, measured in the first month, was associated with employee counterproductive work behavior, measured in consecutive months. Interestingly, the high level of employee counterproductive work behavior reversely led to the intensification of harassing behaviors by the manager. The study results indicate that there is a certain negative spiral gain between these two factors. Similarly, Chinese researchers, in a longitudinal five-fold measurement study, identified a positive relationship between abusive supervision and all dimensions of counterproductive work behavior (Wei and Si 2013). They observed the weakest dependencies in the relationship between abuse and sabotage, and the strongest in the relationship between abuse and theft.

5.1.4 Job Burnout as a Mediator of the Relationship between Social Stressors and Counterproductive Work Behavior

Job burnout is defined as a result of excessive job demands and lack of sufficient resources to cope with these requirements (Demerouti et al. 2001). It comprises two components: *exhaustion* and *disengagement from work*. The authors describe exhaustion as a result of persistent, chronic tension caused by physical, emotional and cognitive job demands. Hence, the researchers put emphasis not only on the emotional but also on the physical and cognitive aspects of exhaustion. Disengagement from work is described as a withdrawing attitude in relation to clients, co-workers, work tasks and the entire working environment, e.g. professional duties, organizational values and culture.

The mediating role of job burnout in the relationship between work stressors and counterproductive work behavior has not been the subject of wider research. Several studies have shown that job burnout mediates the effect of stressors on employee counterproductive work behaviors, such as absenteeism (Bakker et al. 2003), low performance (Bakker et al. 2004) and disrespectful handling of clients (van Jaarsveld et al. 2010). Two other studies have supported the mediating role of job burnout in the relationship between insufficient workload and counterproductive work behaviors (Luksyte et al. 2011), and in the relationship between excessive workload and counterproductive work behaviors (Smoktunowicz et al. 2015).

Other studies have shown that job burnout is a form of self-defense, a protective mechanism in coping with excessive job demands, such as in special character professions involving close and emotionally exhausting relations with patients or beneficiaries of support (Cordes and Dougherty 1993). Maslach et al. (2001), among techniques used by employees to distance themselves from job demands, has identified the use of professional jargon, intellectualization, maintaining a clear work/life boundary, humor and withdrawal of work engagement. More recent studies have revealed that distanced employees have much worse attitudes toward the entire workplace context, e.g. they have lower work engagement levels (Banks et al. 2012), create negative categorizations and more often perceive their co-workers as "the other" (Bolton et al. 2012). Counterproductive work behaviors may also appear as a consequence of the distanced, negative attitude toward work. It is worth mentioning here the conservation of resources theory (Hobfoll 2006). It implies that workers with high levels of exhaustion and low levels of resources should be strongly motivated to manage the resources rationally – to save the remaining resources and to recover the resource already used. This can be achieved by reducing work engagement and performance levels, avoiding a high workload, frequent absenteeism, work delays and extended breaks. For example, an intentional decline in performance levels can contribute to energy saving, while prolonged rest breaks and absenteeism can help in recovering physical and mental strength (Krischer et al. 2010; Wilson 2015).

It thus seems that job burnout plays an important role in the emergence of negative behaviors at work. The evidence has supported the direct relationship between job burnout and various types of counterproductive work behaviors (e.g. Banks et al. 2012; Bolton et al. 2012; Liang and Hsieh 2007; Leiter and Robichaud 1997), as well as the mediating role of job burnout in the relationship between work stressors and counterproductive work behaviors (Luksyte et al. 2011; Smoktunowicz et al. 2015; van Jaarsveld et al. 2010). However it remains unclear how the individual job burnout components mediate the relationship between work stressors and counterproductive work behavior. Accordingly, the mediating role of two job burnout components (i.e. exhaustion and disengagement from work) is the subject of the present study.

5.1.5 The Moderating Role of Job Control and Social Support at Work

Following the *stressoremotion model* (Spector and Fox 2005), employees with strong job control less often engage in counterproductive work behaviors when confronted with work stressor experiences. This premise stems from the tradition of occupational stress research, conducted in the context of the *job demands–control model* (Karasek 1979), whereby job control has been identified as a stress buffer. Further research into the role of job control has shown that it does indeed play a beneficial role, in particular when combined with social support (Karasek and Theorell 1990; Häusser et al. 2010). The cumulative role of job control and social support has been emphasized in many stress concepts, including the *demands–control–support* model (Karasek and Theorell 1990), the conservation of resources theory (Hobfoll 2006) and the *job demands–resources model* (Bakker et al. 2003). At a general level, these concepts presume that the negative effects of stress are a result of the combined effects of high job demands, low job control and low social support. In the vast majority of studies, job control and social support have been analyzed in the context of occupational health. Since the primary sources of CWB are work stressors, it is worth examining whether these resources also reduce the level of CWB.

The authors of the stressor-emotion model have built on a premise that the very awareness of job control, i.e. the freedom of action, the possibility of independent decision making, testing new solutions and taking responsibility for achieved results, may empower employees to cope more effectively with stressful situations, and experience a lower number of negative emotions. Nevertheless, this concept has not been empirically confirmed. In two known studies, the moderating effect of job control in the relationship between stressor and counterproductive work behavior has been tested (Fox et al. 2001; Tucker et al. 2009). Both study results are inconsistent with the stressor-emotion model (Spector and Fox 2005). One of the studies has demonstrated that job control intensifies (and does not buffer) the effect of interpersonal conflicts on counter-productive behaviors (Fox et al. 2001). Similar results have been obtained by Tucker et al. in longitudinal, six-month measurement interval studies in a group of peacekeeping mission soldiers, including soldiers deployed in Kosovo and Kuwait missions. When confronted with excessive duty service demands, soldiers with the greatest job control most often violated military discipline and regulations (Tucker et al. 2009). These data have shown that job control, in many theories perceived as a stress buffer, may in fact exacerbate stress in some cases.

As for the moderating role of social support, a recent study by Chinese researchers has demonstrated that support from managers and co-workers buffers the negative effect of work stressors (i.e. conflict, role uncertainty and work overload) on counterproductive work behavior (Chiu et al. 2015). Also, a Polish study has proved that a high level of social support at work mitigates the direct association between workload and counterproductive work behavior (Smoktunowicz et al. 2015). However, other researchers have not confirmed the moderating role of social support at work in the relationship between negative emotions and counterproductive work behaviors (Sakurai and Jex 2012).

The aim of the present research is to investigate the direct relationship between two types of social stressors (i.e. interpersonal conflicts at work and workplace bullying) and to determine how two components of job burnout (i.e. exhaustion and disengagement from work) mediate this relationship. The moderating role of job control and social support will also be examined. The research hypotheses are outlined below:

H1: Social stressors are positively associated with counterproductive work behavior.

H2: Exhaustion mediates the effect of social stressors on counterproductive work behavior.

H3: Disengagement from work mediates the effect of social stressors on counterproductive work behavior.

H4: Job resources buffer the effect of social stressors on counterproductive work behaviors.

5.2 STUDY METHOD

5.2.1 Study Sample and Procedure

The survey was conducted on a sample of 2,597 police officers. Men made up 81% of the study sample (2,102 persons), women – 19% (496 persons). The age of the respondents was from 20 to 63 years (*M* = 38.92, *SD* = 9.23). The career length ranged from 1 to 36 years (*M* = 13.41; *SD* = 8.06). The research was conducted by interviewers using the PAPI technique in police stations all over Poland. The research was voluntary and anonymous. Each police officer received a set of questionnaires from a trained interviewer and filled it out in a room prepared for this purpose. The completed questionnaire sheets were returned to the interviewers in sealed envelopes.

5.2.2 Measurement of the Study Variables

Social stressors. Interpersonal conflicts were measured using Interpersonal Conflicts at Work Scale (ICAWS) (Spector and Jex 1998). The ICAWS contains four items (e.g. *How often do you argue with others at work? How often do people at work yell at you?*), with a five-point answer scale (from 1 – *less than once a month or never*, to 5 – *several times a day*). The reliability index of the scale was Cronbach's α = 0.87. The measurement of workplace bullying was based on the results of the *Negative Acts Questionnaire* (Einarsen and Raknes 1997). This tool consists of 22 items measuring the frequency of experiencing different negative behaviors at work, considered as instances of workplace bullying (e.g. *How often in the last six months have you had to ignore your ideas and views?*). Answers are provided on a five-point scale (from 1 – *never* to 5 – *daily*). The questionnaire can be divided into two subscales – workplace bullying targeting the personal sphere and workplace bullying directed at the professional sphere. In the study, the aggregate bullying index was taken into account. The reliability of the tool used in the study was high (Cronbach's α = 0.94).

Job resources. Job control and social support were measured by two subscales that are part of the *Job Content Questionnaire* (Karasek 1985). The job control measuring scale consists of nine items, covering the level of employee autonomy and the degree of freedom of action while performing tasks (e.g. *My work allows me to make my own decisions*) and the degree of skill use related to the level of work complexity (e.g. *My work requires performing various tasks and activities*). The social support scale includes nine items. Four items concern subjective evaluation of the possibility of obtaining help from managers; the five remaining items relate to subjective evaluation of the possibility of obtaining help from co-workers. All the items that make up the job control and social support subscales are answered on a four-point scale (from 1 – *I strongly disagree*, to 4 – *I strongly agree*). In the present study, aggregate indicators of job control and social support have been considered. The reliability coefficients (Cronbach's α) for job control and social support scales are 0.76 and 0.85 respectively.

Job burnout. The variable was measured with the *Oldenburg Burnout Inventory* (Demerouti et al. 2001). The questionnaire contains 16 items which consist of two dimensions of burnout: *exhaustion* and *disengagement from work*. For each of the subscales there are eight statements, including four with a reverse coding of results. The reliability coefficients for the two subscales are 0.82 for the exhaustion, and 0.76 for the disengagement from work.

Counterproductive work behavior. To measure this variable, an abbreviated version of the *Counterproductive Work Behavior Questionnaire* (Checklist, CWB-C) developed by Spector et al. (2006) was used. It contains 32 items, with a five-point answer scale (from 1 – *never*, to 5 – *every day*). The Cronbach's reliability index was α = 0.92.

5.2.3 Analytical Procedures

A Bonferroni's correction (α = 0.004) reliability analysis and *r-Pearson* correlation analysis of the study variables were carried out. The correlation analysis, in addition to the main study variables, also included demographic variables, i.e. age, gender and career length. Direct predictor effects, as well as mediational and moderating effects were tested using the structural modeling method. The following models were used: a single predictor (stressor) model, two mediating factors' (exhaustion and disengagement from work) model, two moderating factors' (job control and social support) model and a single dependent variable model (counterproductive work behavior). Mediational effects were tested on the basis of standardized regression coefficients values (Beta) of path *a* (predictor effect on the mediator), path *b* (mediator effect on the dependent variable) and $a \times b$ interaction (total mediational effect). Moderation analyses were tested based on the interaction of latent variables. The interaction between the predictors and the moderator was orthogonalized by standardizing the regression analysis residuals, whereby all observable moderator indicators and the given predictor were used to explain the interaction. These interactions were the result of a combination of three observable moderator indicators, and three observable indicators of the analyzed predictor (Little et al. 2006). In order to test the moderating effects, a base latent moderating factor and an interaction latent factor were introduced into the model. The Satorra–Bentler correction (Satorra and Bentler 1988) ML method was used in the analyses.

5.3 STUDY RESULTS AND VERIFICATION OF THE STUDY HYPOTHESES

The study results showed that demographic variables (age and career length) negatively correlated with workplace bullying ($r = -0.10$; $p < 0.05$ and $r = -0.13$; $p < 0.01$). In addition, professional career length was positively associated with job control ($r = 0.14$; $p < 0.01$) and negatively associated with perceived social support ($r = -0.12$; $p < 0.05$). A weak, negative correlation was observed between age, career length and counterproductive work behavior ($r = -0.11$; $p < 0.05$ and $r = -0.08$; $p < 0.05$). Also, a weak correlation between gender and counterproductive work behavior ($r = -0.10$; $p < 0.05$) was observed. The comparative Student's t-test analysis showed that women engaged less frequently than men in activities detrimental to the organization ($t = -3.05$; $M = 36.94 < M = 38.37$).

Table 5.2 outlines the study variables' correlation analysis results. It shows that counterproductive work behaviors are positively associated with two social stressors and correlated negatively with two types of job resources – job control and social support. The analyzed stressors are positively associated with the two components of job burnout –exhaustion and disengagement from work. These components are negatively associated with job control and social support. Exhaustion and disengagement from work reveal positive associations with counterproductive work behaviors.

The study hypotheses were examined on the basis of structural modeling results. The tested models included a predictor (social stressor), two mediators (exhaustion and disengagement from work) and two moderators (job control and social support). Data fit measures of the two tested models (χ^2, *df*, RMSEA, CFI, SRMR) are presented in Table 5.3. They indicate a good model to data fit. The **H1** held that social stressors would have been directly associated with counterproductive work behavior. Figures 5.2 and 5.3 show the results proving these dependencies (c' path) in relation to interpersonal

TABLE 5.2

Correlation Matrix, Variable Mean and Standard Deviation Values

	(1)	(2)	(3)	(4)	(5)	(6)	M	SD
(1) Interpersonal conflict	–						1.92	0.81
(2) Bullying	0.54***						2.07	0.81
(3) Exhaustion	0.30***	0.34***					2.54	0.61
(4) Disengagement from work	0.29***	0.37***	0.64***				2.48	0.59
(5) Job control	–0.22***	–0.28***	–0.26***	–0.34***			3.88	0.45
(6) Social support	–0.39***	–0.39***	–0.39***	–0.44***	0.32***		2.33	0.39
(7) CWB	0.38***	0.31***	0.11**	0.16***	–0.14***	–0.10**	1.61	0.40

* $p < 0.05$; **$p < 0.01$; ***$p < 0.001$.

$p < 0.05$ and $p < 0.01$ Bonferroni correction correlation levels are statistically insignificant.

TABLE 5.3
Structural Models' Parameters of the Study

Model	χ^2	*df*	RMSEA	90% CI	CFI	SRMR
Model 1: ICAW → CWB	5662.966	478	0.047	0.044; 0.050	0.918	0.017
Model 2: Bullying → CWB	5365.212	478	0.054	0.046; 0.058	0.922	0.021

Note: ICAW – interpersonal conflict at work, CWB – counterproductive work behavior. All χ^2 values are significant at the $p < 0.001$ level. Each of the above models includes one predictor (stressor), two mediating factors (exhaustion and disengagement from work), two moderating factors (job control and social support) and a dependent variable (counterproductive work behavior).

conflicts (Figure 5.2, $\beta = 0.668$; $p < 0.001$) and workplace bullying (Figure 5.3, $\beta = 0.554$; $p < 0.001$). The results indicate that both types of social stressors are predictive of counterproductive work behaviors. The **H1** has been fully confirmed.

It was presumed that exhaustion (H2) and disengagement from work (H3) would mediate the relationship between social stressors and counterproductive work behavior. The analysis showed that disengagement from work (Figure 5.2 *a1* and *b1* paths),

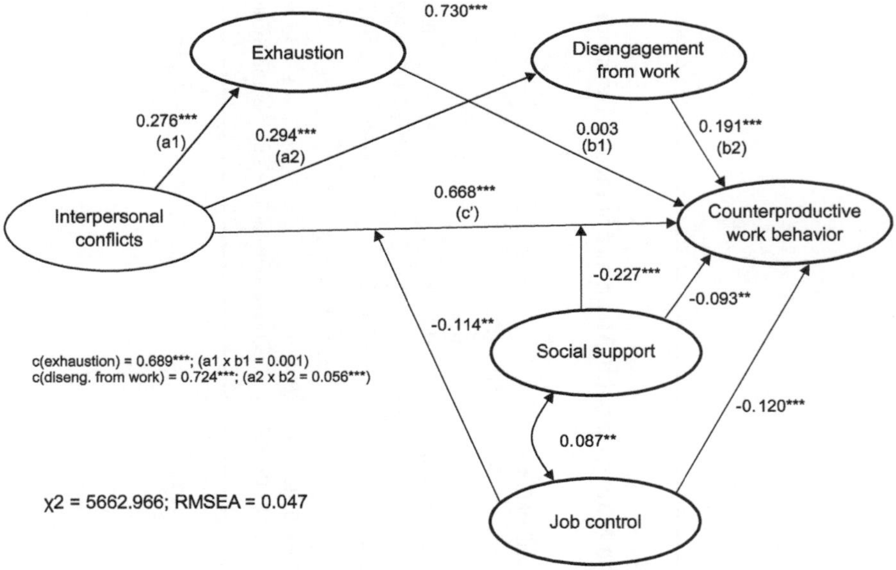

FIGURE 5.2 Effect of interpersonal conflicts on counterproductive work behaviors, mediating role of exhaustion and disengagement from work, with job control and social support as moderating factors.

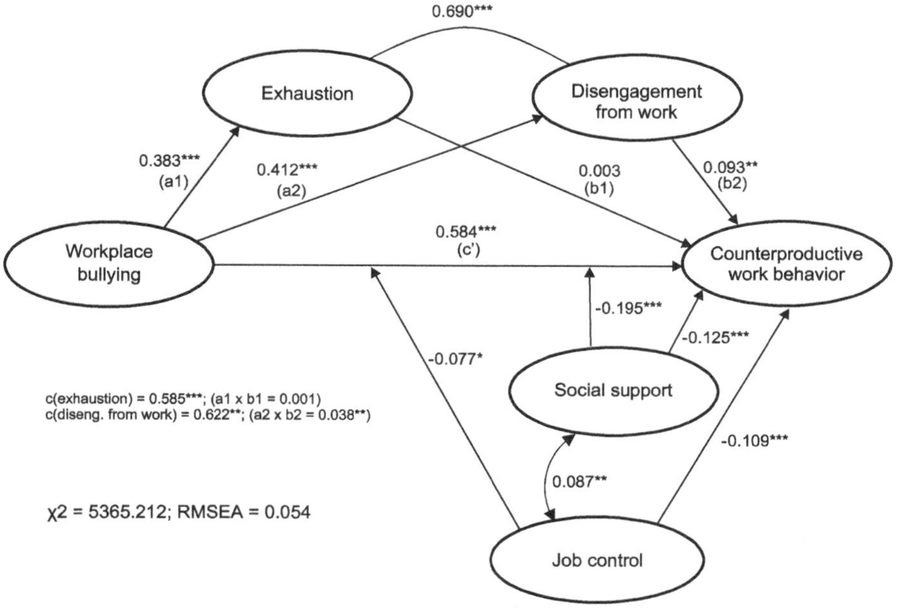

FIGURE 5.3 Effect of workplace bullying on counterproductive work behaviors, mediating role of exhaustion and disengagement from work, with job control and social support as moderating factors.

but not exhaustion (Figure 5.3, *a1* and *b1* path), was a mediator of the negative effect of the two types of social stressors on CWB. The obtained results have not confirmed the **H2**, but have fully supported the **H3**. The role of job resources (including job control and social support) in the development of CWB has also been analyzed. The obtained results proved that both job control and social support were predictors of a low level of CWB. The moderating effect of job resources (**H4**) has also been confirmed. The analyses showed that both job control and social support mitigated the negative effects of interpersonal conflicts (Figure 5.2; $\beta = -0.114$; $p < 0.01$ and $\beta = -0.227$; $p < 0.001$, respectively) and workplace bullying (Figure 5.3; $\beta = -0.077$; $p < 0.05$ and $\beta = -0.195$, $p < 0.001$, respectively) on CWB.

5.4 DISCUSSION

The study has examined the direct relationship between two types of social stressors at work and counterproductive work behaviors, and the indirect relationship mediated by two components of job burnout, i.e. exhaustion and disengagement from work. The moderating role of job control and social support has also been investigated. The analysis results have proved that both interpersonal conflicts at work and workplace bullying are positively associated with counterproductive work behaviors. These results are consistent with the *retaliation model* (e.g. Skarlicki and Folger 1997) and the *revenge model* (Bies and Tripp 2005). According to these models, counterproductive work behavior is motivated by the desire to retaliate or take revenge for bad treatment suffered at work, and is accompanied by strong emotions of frustration, anger and hostility. Social stressors are the source of such emotions.

Regarding the mediational effects, the research has only confirmed the mediating role of disengagement from work (but not exhaustion). A high level of social stressors at work, conducive to increased disengagement from work, exacerbated dysfunctional behaviors in the organization. The obtained data have been consistent to some extent with previous research findings (e.g. Bolton et al. 2012; Leiter and Robichaud 1997; Liang and Hsieh 2007; Luksyte et al. 2011). Although those authors have used earlier tools for testing burnout, and thus have considered other components of job burnout, they have shown that depersonalization and cynicism (i.e. equivalents of disengagement from work) are stronger determinants of dysfunctional organizational behaviors than emotional exhaustion (measured by MBI) and exhaustion (measured by MBI-GS). Nevertheless, an OLBI study investigating the mediating role of job burnout in the relationship between social stressors and counterproductive work behavior does not seem to have been conducted yet. Given that researchers have increasingly been using this scale to measure job burnout (Halbesleben and Demerouti 2010) as an alternative to other tools over the past decade, the results seem to be important in terms of further research on the relationship between job burnout and organizational behavior.

The analysis has also shown that both job control and social support at work are moderating factors in the relationship between social stressors and counterproductive work behaviors. The results are consistent with the *demands–control–support model* (Karasek and Theorell 1990). Although this model has been tested mainly in the area of health (e.g. Hausner et al. 2010), it seems to be a reliable model in the area of organizational behaviors. In subsequent studies, it is worth exploring the role of personality in the development of these behaviors. The effect of the interaction between organizational factors and personality traits seems to be particularly interesting, and it has been observed by several authors, including Folger and Skarlicki (Folger 1998).

The presented research results demonstrate that social stressors in the working environment have a negative effect not only on employee health but also on organizational behavior. The costs of these behaviors are high and calculated at both enterprise and country levels. A way to reduce these costs could be the implementation of prevention programs aimed at assisting employees in coping with work-related stress. Such programs have successfully been rolled out in the United States and in Western European countries – mainly in Scandinavia, Germany and the Netherlands. The prevention schemes include both individual training to prepare employees to actively reduce cognitive and behavioral stress, as well as management workshops to raise the management awareness of the negative consequences of stress, and to show effective ways of implementing employee-friendly measures in the organization. Another way to help employees cope with stress is to build organizational resources such as respecting the psychological contract, creating a social support structure between employees and management, strengthening employee job control and providing employees with feedback on their performance. Many studies have shown that robust organizational resources can support employees in more efficient coping with work-related stress (e.g. Widerszal-Bazyl 2003).

Finally, it is worth highlighting the need for research on the causes, mechanisms and effects of counterproductive work behaviors, also in Poland. This is particularly motivated by the paucity of studies and practical knowledge on the emergence of counterproductive work behaviors in Polish enterprises. Longitudinal studies would be particularly valuable. Counterproductive work behavior is a highly dynamic phenomenon and develops as a result of long-term effects and emotions associated with social stressors and job resources. To have a full grasp of these relationships, cross-lagged studies with measurement intervals of a few months would be recommended.

REFERENCES

Adams, J. S. 1965. Inequity in social exchange. In: *Advances in Experimental Social Psychology*, ed. L. Berkowitz, vol. 2, 267–299. New York: Academic Press.

Bakker, A. B., E. Demerouti, E. De Boer, and W. B. Schaufeli. 2003. Job demands and job resources as predictors of absence duration and frequency. *J Vocat Behav* 62(2):341–356.

Bakker, A. B., E. Demerouti, and W. Verbeke. 2004. Using the Job Demands-Resources model to predict burnout and performance. *Hum Resour Manage* 43(1):83–104.

Balducci, C., W. B. Schaufeli, and F. Fraccaroli. 2011. The job demands-resources model and counterproductive work behaviour: The role of job-related affect. *Eur J Work Organ Psychol* 20(4):467–496.

Banks, G. C., C. E. Whelpley, and O. In-Sue. 2012. (How) are emotionally exhausted employees harmful? *Int J Stress Manag* 19(3):198–216.

Bauer, J. A., and P. E. Spector. 2015. Discrete negative emotions and counterproductive work behavior. *Hum Perform* 28(4):307–331.

Berry, C. M., N. C. Carpenter, and C. L. Barratt. 2012. Do other-reports of counterproductive work behavior provide an incremental contribution over self-reports? A meta-analytic comparison. *J Appl Psychol* 97(3):613–636.

Bennett, R. J., and S. L. Robinson. 2000. Development of a measure of workplace deviance. *J Appl Psychol* 85(3):349–360.

Bibi, Z., and J. Karim. 2013. Workplace incivility and counterproductive work behavior: Moderating role of emotional intelligence. *Pak J Psychol Res* 28(2):317–334.

Bies, R. J., and T. M. Tripp. 1998. The many faces of revenge. The good, the bad, and the ugly. In: *Dysfunctional Behavior in Organizations, Part B: Nonviolent Dysfunctional Behavior*, eds. R. W. Griffin, A. O'Leary-Kelly and J. M. Collins, 49–68. Stanford: JAI Press.

Bies, R. J., and T. M. Tripp. 2005. The study of revenge in the workplace: Conceptual, ideological, and empirical issues. In: *Counterproductive Work Behavior: Investigations of Actors and Targets*, eds. S. Fox and P. E. Spector, 65–81. Washington, DC: American Psychological Association.

Bies, R. J., T. M. Tripp, and R. M. Kramer. 1997. At the breaking point: Cognitive and social dynamics of revenge in organizations. In: *Antisocial Behavior in Organizations*, ed. R. A. Giacalone and J. Greenberg, 18–36. Thousand Oaks, CA: Sage Publications.

Bolton, L. R., R. D. Harvey, M. J. Grawitch, and L. K. Barber. 2012. Counterproductive work behaviours in response to emotional exhaustion: A moderated mediational approach. *Stress Health* 28(3):222–233.

Bruk-Lee, V., and P. Spector. 2006. The social stressors – Counterproductive work behaviors link: Are conflicts with supervisors and coworkers the same? *J Occup Health Psychol* 11(2):145–156.

Chiu, S. F., S. P. Yeh, and T. C. Huang. 2015. Role stressors and employee deviance: The moderating effect of social support. *Pers Rev* 44(2):308–324.

Cordes, C., and T. W. Dougherty. 1993. A review and integration of research on job burnout. *Acad Manag Rev* 18(4):621–656.
Demerouti, E., A. B. Bakker, F. Nachreiner, and W. B. Schaufeli. 2001. The job demands-resources model of burnout. *J Appl Psychol* 86(3):499–512.
Einarsen, S., H. Hoel, D. Zapf, and C. L. Cooper. 2003. The concept of bullying at work. In: *Bullying and Emotional Abuse in the Workplace: International Perspectives in Research and Practice*, eds. S. Einarsen, H. Hoel and D. Zapf, 3–30. London: Taylor & Francis.
Einarsen, S., and B. Raknes. 1997. Harassment in the workplace and the victimization of men. *Violence Vict* 12(3):247–263.
Folger, R., and D. P. Skarlicki. 1998. A popcorn metaphor for employee aggression. In: *Dysfunctional Behavior in Organizations: Violent and Deviant Behavior*, eds. R. W. Griffin, A. O'Leary-Kelly and J. M. Collins, 43–81. Stanford: JAI Press.
Folger, R., and D. Skarlicki. 2005. Beyond counterproductive work behavior: Moral emotions and deontic retaliation Versus reconciliation. In: *Counterproductive Work Behavior: Investigations of Actors and Targets*, eds. S. Fox and P. Spector, 83–105. Washington, DC: American Psychological Association.
Fox, S., P. E. Spector, and D. Miles. 2001. Counterproductive work behavior (CWB) in response to job stressors and organizational justice: Some mediator and moderator tests for autonomy and emotions. *J Vocat Behav* 59(3):291–309.
Giacalone, J., and J. Greenberg. 1997. *Antisocial Behavior in Organizations*. Thousand Oaks, CA: Sage.
Greenberg, J. 1990. Employee theft as a reaction to underpayment inequity: The hidden cost of pay cuts. *J Appl Psychol* 75(5):561–568.
Halbesleben, J. R. B., and E. Demerouti. 2010. The construct validity of an alternative measure of burnout: Investigating the English translation of the Oldenburg Burnout Inventory. *Work Stress* 19(3):208–220.
Häusser, J. A., A. Mojzisch, M. Niesel, and S. Schulz-Hardt. 2010. Ten years on: A review of recent research on the Job Demand-Control-Support model and psychological well-being. *Work Stress* 24(1):1–35.
Hershcovis, M. S., N. Turner, J. Barling et al. 2007. Predicting workplace aggression: A meta-analysis. *J Appl Psychol* 92(1):228–238.
Hobfoll, S. E. 2006. *Stres, kultura i społeczność. Psychologia i filozofia stresu* [Stress, culture and community. Psychology and philosophy of stress]. Gdańsk: GWP.
Hogan, R., and J. Hogan. 1989. How to measure employee reliability. *J Appl Psychol* 74(2):273–279.
Hollinger, R. C., and J. P. Clark. 1983. *Theft by Employees*. Lexington, KY: Lexington Books.
Karasek, R. A. 1979. Job demands, job decision latitude and mental strain: Implications for job redesign. *Admin Sci Q* 24(2):285–308.
Karasek, R. A. 1985. *Job Content Questionnaire and User's Guide*. Lowell, MA: University of Massachusetts.
Karasek, R. A., and T. Theorell. 1990. *Healthy Work. Stress, Productivity and the Reconstruction of Working Life*. New York: Basic Books.
Keenan, A., and T. J. Newton. 1985. Stressful events, stressors and psychological strains in young professional engineers. *J Organ Behav* 6(2):151–156.
Kessler, S. R., K. Bruursema, B. Rodopman, and P. E. Spector. 2013. Leadership, interpersonal conflict, and counterproductive work behavior: An examination of the stressor/strain process. *Negot Confl Manag Res* 16:180–190.
Krischer, M. M., L. M. Penney, and E. M. Hunter. 2010. Can counterproductive work behaviors be productive? CWB as emotion – Focused coping. *J Occup Health Psychol* 15(2):154–166.

Leiter, M. P., and L. Robichaud. 1997. Relationships of occupational hazards with burnout: An assessment of measures and models. *J Occup Health Psychol* 2(1):35–44.
Liang, S. C., and A. T. Hsieh. 2007. Burnout and workplace deviance among flight attendants in Taiwan. *Psychol Rep* 101(2):457–468.
Little, T. D., J. A. Bovaird, and K. F. Widaman. 2006. On the merits of orthogonalizing powered and product terms: Implications for modeling interactions among latent variables. *Struct Equ Model* 13(4):497–519.
Liu, C., M. M. Nauta, C. P. Li, and J. Y. Fan. 2010. Comparisons of organizational constraints and their relations to strain in China and the United States. *J Occup Health Psychol* 15(4):452–467.
Luksyte, A., C. Spitzmueller, and D. C. Maynard. 2011. Why do overqualified incumbents deviate? Examining multiple mediators. *J Occup Health Psychol* 16(3):279–296.
Maslach, C., W. B. Schaufeli, and M. P. Leiter. 2001. Job burnout. *Annu Rev Psychol* 52:397–422.
Mitchell, M. S., and M. L. Ambrose. 2007. Abusive supervision and workplace deviance and the moderating effects of negative reciprocity beliefs. *J Appl Psychol* 92(4):1159–1168.
Narayanan, L., S. Menon, and P. Spector. 1999. A cross-cultural comparison of job stressors and reaction among employees holding comparable jobs in two countries. *Int J Stress Manag* 6(3):197–212.
Neuman, J. H., and R. A. Baron. 1998. Workplace violence and workplace aggression: Evidence concerning specific forms, potential causes, and preferred targets. *J Manag* 24(3):391–419.
Penney, L. M., and P. E. Spector. 2007. Emotions and counterproductive work behavior. In: *Research Companion to Emotion in Organizations*, eds. N. M. Ashkanasy and C. L. Cooper, 183–196. Northampton, MA: Edward Elgar Publishing.
Robinson, S. L., and R. J. Bennett. 1995. A typology of deviant workplace behaviors: A multidimensional scaling study. *Acad Manag J* 38(2):555–572.
Sakurai, K., and S. M. Jex. 2012. Coworker incivility and incivility targets' work effort and counterproductive work behaviors. The moderating role of supervisor social support. *J Occup Health Psychol* 17(2):150–161.
Satorra, A., and P. Bentler. 1988. *Scaling Corrections for Statistics in Covariance Structure Analysis*. Los Angeles, CA: Department of Statistics, UCLA.
Simon, L. S., C. Hurts, K. Kelley, and T. Judge. 2015. Understanding cycles of abuse. A multimotive approach. *J Appl Psychol* 100(6):1798–1810.
Skarlicki, D. P., and R. Folger. 1997. Retaliation in the workplace: The roles of distributive, procedural, and interactional justice. *J Appl Psychol* 82(3):434–443.
Smoktunowicz, E., L. Baka, R. Cieslak, C. F. Nichols, C. C. Benight, and A. Luszczynska. 2015. Explaining counterproductive work behaviors Among police officers: The indirect effects of job demands are mediated by burnout and moderated by job control and social support. *Hum Perform* 28(4):332–350.
Spector, P. E., and S. Fox. 2005. *Counterproductive Work Behavior: Investigations of Actors and Targets*. Washington, DC: American Psychological Association.
Spector, P. E., S. Fox, L. M. Penney, K. Bruursema, A. Goh, and S. Kessler. 2006. The dimensionality of counterproductivity: Are all counterproductive behaviors created equal? *J Voc Beh* 68(3):446–460.
Spector, P. E., and S. M. Jex. 1998. Development of four self-report measures of job stressors and strain: Interpersonal conflict at work scale, organizational constraints scale, quantitative workload inventory and physical symptoms inventory. *J Occup Health Psychol* 3(4):356–367.
Tepper, B. J. 2000. Consequences of abusive supervision. *Acad Manag J* 43(2):178–190.

Tripp, T. M., and R. J. Bies. 1997. What's good about revenge? The avenger's perspective. In: *Research on Negotiation in Organizations*, eds. R. J. Lewicki, R. J. Bies and B. H. Sheppard. Greenwich: JAI Press.

Tripp, T. M., R. J. Bies, and K. Aquino. 2002. Poetic justice or petty jealousy? The aesthetics of revenge. *Organ Behav Hum Decis Process* 89(1):966–984.

Tucker, J. S., R. R. Sinclair, C. D. Mohr, J. L. Thomas, A. D. Salvi, and A. B. Adler. 2009. Stress and counterproductive work behavior: Multiple relationships between demands, control, and soldier indiscipline over time. *J Occup Health Psychol* 14(3):257–271.

van Jaarsveld, D. D., D. D. Walker, and D. P. Skarlicki. 2010. The role of job demands and emotional exhaustion in the relationship between customer and employee incivility. *J Manag* 36(6):1486–1504.

Warszewska-Makuch, M. 2013. Osobowościowe i sytuacyjne predyktory mobbingu w miejscu pracy i jego relacje z samopoczuciem psychicznym i satysfakcją z pracy [Personal and situational predictions of workplace mobbing and its relationship to mental well-being and job satisfaction]. Unpublished doctoral thesis. Warszawa: SWPS. Faculty of Psychology.

Wei, F., and S. Si. 2013. Tit for tat? Abusive supervision and counterproductive work behaviors. The moderating effects of locus of control and perceived mobility. *Asian J Manag* 30(1):281–296.

Widerszal-Bazyl, M. 2003. *Stres w pracy a zdrowie czyli o próbach weryfikacji modelu Roberta Karaska oraz modelu: Wymagania – Kontrola – Wsparcie* [Stress at work and health - about attempts to verify robert Karasek's model and the requirements - control - support model]. Warszawa: Central Institute for Labour Protection.

Wilson, R. A., S. J. Perry, L. A. Witt, and R. W. Griffeth. 2015. The exhausted short-timer: Leveraging autonomy to engage in production deviance. *Hum Relat* 68(11):1693–1711.

6 Workplace Bullying, Mental Health and Job Satisfaction

The Moderating Role of the Individual Coping Style

Magdalena Warszewska-Makuch

CONTENTS

6.1 INTRODUCTION

6.1.1 Workplace Bullying

Most researchers identify workplace bullying as a form of extreme social stress (Niedl 1995; Zapf 1999). Hoel et al. (1999), based on the classical categorization of stressors at work developed by Cooper and Marshall (1976), includes workplace bullying in the group termed "relations at work". Workplace intimidation refers to exposure to systematic and persistent adverse activities in relation to which the worker feels helpless and has little opportunity to defend himself/herself (Einarsen et al. 2009; Einarsen and Raknes1997). The negative behaviors are primarily psychological in nature, but can also be manifested directly in physical violence or threats of physical violence (Einarsen et al. 2009).

In the case of workplace bullying there is no single, universal definition. One of the most frequently quoted definitions of workplace bullying, which was created as a result of the joint work of researchers from Norway, Germany and the United Kingdom, is that of Einarsen, Hoel, Zapf and Cooper (Einarsen et al. 2003). Workplace bullying is defined here as

> harassing, offending, socially, excluding someone, or negatively affecting someone's work. In order for the label bullying (or mobbing) to be applied to a particular activity, interaction or process it has to occur repeatedly and regularly (e.g. weekly) and over a period of time (e.g. about six months). Bullying is an escalating process in the course of which the person confronted ends up in an inferior position and becomes the target of systematic negative social acts. A conflict cannot be called workplace bullying if the incident is an isolated event or if two parties of approximately equal "strength" are in conflict.

When defining workplace bullying, the authors put particular emphasis on two factors, i.e. repeatability and persistence of the phenomenon (Einarsen et al. 2003; Leymann 1990; Zapf et al. 1996). This means that workplace bullying is not a single, isolated act, but rather a series of activities that over a longer period of time are regularly directed against the selected victim/s. Workplace bullying does not constitute a homogeneous behavior and seems to be based on a variety of activities, ranging from such subtle and difficult to recognize forms as gossiping behind the victim's back, or skipping the victim when sending out e-mails with relevant information, to drastic acts involving physical violence. At the same time, it should be stressed that the whole spectrum of undesirable, negative behaviors has one goal in common, i.e. to humiliate, intimidate or punish the person(s) they are aimed at (Einarsen et al. 2003).

6.1.2 Workplace Bullying, Mental Health and Job Satisfaction

A number of studies show that workplace bullying is associated with experiencing strong, prolonged stress leading to serious mental and physical health disorders (Einarsen and Raknes1997; Hoel et al. 1999; Vartia 2001; Zapf et al. 1996; Notelaers et al. 2006; Skogstad et al. 2007), including post-traumatic stress disorder (PTSD) (Matthiesen and Einarsen 2004; Mikkelsen and Einarsen 2002; Tehrani 2004), anxiety and depression (Kivimaki et al. 2003). As a result, such people are no longer able to work and take early retirement or disability pensions. In O'Moore's study (2000), 40% of the victims surveyed admitted that workplace bullying affected their physical health and their mental health (43%). Correspondingly, 26% of respondents with physical health disorders and 92% of respondents with mental health disorders consulted a specialist with workplace bullying problems. One in five respondents declared that as a result of these traumatic experiences they were treated pharmacologically. Leymann (1992), on the other hand, based on the results of a study conducted in a group of Swedish workers, concluded that the strongest distinction could be made between bullies and non-bullied persons in terms of the "cognitive effects" of workplace bullying (annoyance, aggression, memory and concentration problems) and psychosomatic symptoms.

The studies conducted so far clearly indicate that job satisfaction is related to worker health and life satisfaction (Zalewska 2006; Zalewska 2003). Job satisfaction refers to the question of to what extent a worker likes his/her work and satisfies his needs and aspirations at work (Zalewska 1999). Many studies show that job satisfaction is connected with many aspects of functioning of workers. This variable is associated with the quality of life, stress, accidents at work (Fraser and Gilliam 1987), health and absence (Fraser and Gilliam 1987; Herzberg et al. 1959), staff turnover and performance (Herzberg et al. 1959; Mikes and Hulin 1968). Poor job satisfaction can also affect loyalty to the organization and entail anti-social behaviors that cause damage to the organization. Nielsen and Einarsen (Nielsen and Einarsen 2012) emphasize that the inability to deal with unwelcome, unfair treatment at work, i.e. workplace bullying, can lead to a significant reduction in job satisfaction, lack of work engagement and even an intention to quit the job. Regarding the job satisfaction alone, empirical studies show a negative relationship between this variable and workplace bullying (Quine 2001; Bowling and Beehr 2006; Einarsen et al. 1998). Most of these studies are cross-sectional, which does not allow a causal interpretation of the results. Among the few longitudinal studies that focus on this relationship are those conducted by Nielsen et al. (2008). Over the course of two years, the authors examined twice more than 1,700 people and proved a significant relationship between experienced workplace bullying and reduced job satisfaction. Similarly Tepper et al. (2004) found that workplace bullying by a supervisor declared in the first measurement had a negative impact on job satisfaction measured seven months later. Longitudinal studies were also carried out by Rodriguez-Munoz et al. (2009) who showed that workplace bullying was an important predictor of job satisfaction.

6.1.3 Ways of Coping with Workplace Bullying as a Form of Extreme Stress

Although workplace bullying has been associated with significant psychological costs (Hoel et al. 2011; Mikkelsen and Einarsen 2002), the research into mechanisms governing this relationship has been initiated only recently (Nielsen and Einarsen 2012). Some studies suggest that the experience of workplace bullying distorts mental functioning as a result of ineffective coping strategies (Lee and Brotheridge 2006), anxiety, prolonged recovery due to excessive fatigue (Rodríguez-Muñoz et al. 2011), and negative affect (Djurkovic et al. 2004; Mikkelsen and Einarsen 2002). For instance, Lee and Brotheridge (Lee and Brotheridge 2006) revealed that the exposure to workplace bullying was associated with ineffective coping strategies of workplace bullying victims (self-blame and lowered self-esteem), which in turn increased the level of job burnout (emotional exhaustion, cynicism, perceived low professional performance), as well as contributing to poor physical health and negative emotions. In addition, Glasø, Bele, Nielsen and Einarsen (Glasø et al. 2011) showed that workplace bullying was an important predictor of leaving work, and the relationship was mediated by the level of work engagement and job satisfaction. Although recent research has provided a valuable insight into the processes underlying the relationship between workplace bullying and individual functioning of workers experiencing workplace intimidation, the dynamics are yet to be explored. This seems to be particularly important as most of the variables studied so far (e.g. job satisfaction and work engagement) have been defined as a consequence of workplace bullying rather than intermediary factors (mediators or moderators) between workplace bullying and the mental functioning of workers (mental health status, affectivity) (Hogh et al. 2011). Knowledge of workplace bullying victims' responses to stressful experiences at work is vital in terms of appropriate intervention strategies to counteract the intimidating practices. An early intervention is fundamental since workplace bullying constitutes an experience beyond the victim's control, who is left with limited options for coping with the suffered intimidation.

From the workplace bullying victim perspective, knowledge of common responses to workplace violence can be helpful in itself, both by normalizing these responses and by facilitating help-seeking behavior. Moreover, when planning workplace bullying counteracting measures, the awareness of bullying as a phenomenon that can affect the well-being of all workers exposed to these negative behaviors, regardless of their personal strengths and coping resources, remains equally important (Raknes et al. 2016).

Previous research has shown that victims of long-term workplace bullying are less likely to use problem-addressing strategies and fail to attempt to resolve the critical issue by taking no action. This may indicate that the victim is deprived of resources to such an extent that he or she is unable to meet the daily job demands. A number of studies conducted so far (Hogh et al. 2011; Raknes et al. 2016) have supported this finding. Despite various efforts made by victims to curb the violent practice, the majority have been unable to do so without third-party support (Hogh and Dofradottir 2001; Zapf and Gross 2001). This is undoubtedly linked to the low predictability and controllability of the workplace bullying experience, whereby victims develop low self-esteem and feel helpless, guilty and depressed. Therefore, the present study constitutes an attempt to determine how individual stress coping styles are related to experiences of workplace bullying and moderate the relationship between workplace bullying and the well-being of affected workers.

6.1.4 Workplace Bullying and Psychosocial Working Conditions

Many researchers (cf. Einaresn et al. 2003) emphasize that it would be too simplistic to explain the phenomenon of workplace bullying solely at the victim level and through the perpetrator's personality. Studies show that the organization itself (organizational culture, working conditions, etc.) also, and perhaps above all, plays a vital role. The fact that the quality of the psychosocial work environment is a factor responsible to a large extent for workplace bullying has already been reiterated by Heinz Leymann (1990; 1993), a pioneer of research on workplace bullying. The author assumed that stress and frustration caused by a negative psychosocial work environment could lead to violence, and, according to the frustration–aggression theory (Berkowitz 1989), workplace bullying can develop on the basis of the perpetrator's aggressive behavior resulting from environmental factors (Einarsen 2000). A stressful environment can also indirectly influence the aggression of the bully, i.e. a person under pressure can violate social norms and thus induce aggressive behavior in other group members (Einarsen 2000). Environmental and organizational sources of workplace bullying are classified in a number of different ways. According to Leymann (Einarsen 1999), the following causes are responsible for a significant proportion of workplace bullying cases: (1) problems with work organization (e.g. role conflict, ambiguity of roles, excessive workloads), (2) poor management/leadership style, (3) social position of the victim, (4) negative or hostile social climate, (5) organizational culture that condones or rewards workplace bullying. Salin (1999) divided the possible organizational causes of workplace bullying into three groups, i.e. enabling, motivating and triggering factors. Enabling factors identify those factors that may increase the risk of bullying at work, but cannot directly trigger it, e.g. job control, social climate. Incentives are factors that can make workplace bullying beneficial and "useful" for the perpetrator and are used to, for example, get rid of threatening competition. An example of a factor belonging to this category is a high level of competition between workers in the company. Triggering factors are factors which suddenly and radically increase the danger of workplace bullying, e.g. change of management, restructuring or staff reduction. In turn, Hoel and Salin (Hoel 2003) identified the following groups of factors conducive to workplace bullying: work organization, the changing nature of work, the climate and organizational culture and leadership.

6.1.5 The Present Study

The present study has been focused on the question of individual stress-coping styles and their role in the relationship between workplace bullying, mental health and job satisfaction of workers, i.e. whether a particular stress-coping style moderates the relationship between the experienced workplace bullying and the well-being of workers.

In addition, when examining the aforementioned dependencies, demographic variables were controlled, i.e. age and gender, important life events, as well as selected factors of the psychosocial work environment, i.e. demands, control and development opportunities.

In current approaches to the study of stress, researchers have increasingly emphasized the significance of effective individual stress-coping styles over the actual experience of work-related stress. Individual stress-coping skills seem to play a moderating role in the relationship between stress and health/ill health. Some authors have reiterated that the health effects of stress transactions are rather determined by the activity undertaken to cope with stress, than by the very instance of stressful experience (Ogińska-Bulik and Juczyński 2008). Also, Antonovsky (1995) considers stress management processes central to explaining health risks, and identifies effective coping strategies as determinants of health. Despite the paucity of a clear understanding of the effectiveness of individual stress-coping styles, several studies on stress have shown that problem-focused styles are more effective.

Workplace bullying is considered a process of severe, systematic and persistent stress experiences over a longer period of time, although the intensity may vary throughout the entire cycle (cf. Zapf and Gross 2001; Einarsen et al. 2003). Some observe that workplace bullying is often prompted by a personal conflict that has not been properly addressed, escalating into prolonged violence by one side against the other. It thus seems that in the initial phase of workplace bullying, the victim's response plays a vital role. The actions undertaken by the victim may either inhibit or exacerbate the development of the violent process. Hence, persons using the task-oriented style seem to be more likely, at least at the beginning of the conflict, to confront the perpetrator, while simultaneously looking for constructive methods of resolving the problem, e.g. a request for a transfer to a different department or team. It thus seems that the task-oriented style would buffer, at least to some extent, the negative impact of workplace bullying on worker well-being. In turn, persons representing the emotion- and avoidance-oriented styles may be more inclined to deny the problem and avoid taking action to end the conflict. Therefore, according to the study hypotheses, the emotion- and avoidance-oriented styles would exacerbate the negative effects of the experienced workplace bullying on the well-being of affected workers.

Figure 6.1 is the research model presenting the assumed relationships.

Based on the developed research model, the following hypotheses have been formulated:

- The task-oriented style of coping with stress is a significant moderator in the relationship between experienced workplace bullying and mental health: the more often the task-oriented style is used, the weaker the association between experienced workplace bullying and negative mental health effects.
- The task-oriented style of coping with stress is a significant moderator in the relationship between experienced workplace bullying and job

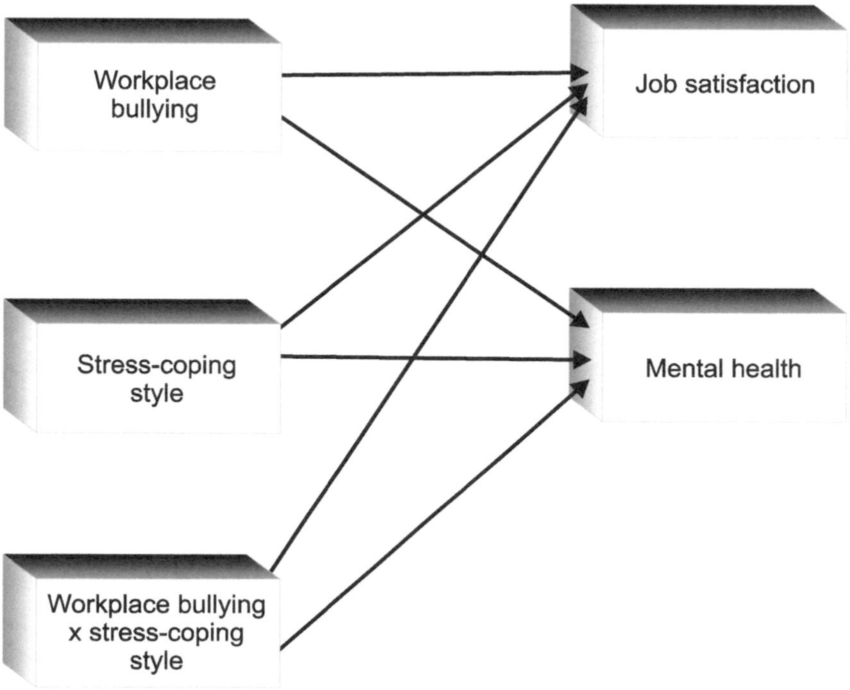

FIGURE 6.1 Research model presenting the assumed relationships between workplace bullying, stress-coping style, mental health and job satisfaction.

satisfaction: the more often the task-oriented style is used, the weaker the relationship between experienced bullying and negative effects on job satisfaction.

- The emotion-oriented style of coping with stress is a significant moderator in the relationship between experienced workplace bullying and mental health: the more often the emotion-oriented style is used, the stronger the association between experienced bullying and negative mental health effects.
- The emotion-oriented style of coping with stress is a significant moderator in the relationship between experienced workplace bullying and job satisfaction: the more often the emotion-oriented style is used, the stronger the association between experienced workplace bullying and negative effects on job satisfaction.
- The avoidance-oriented style of coping with stress is a significant moderator in the relationship between experienced workplace bullying and mental health: the more often the avoidance-oriented style is used, the stronger the association between experienced workplace bullying and negative mental health effects.
- The avoidance-oriented style of coping with stress is a significant moderator between experienced workplace bullying and job satisfaction: the

more often the avoidance-oriented style is used, the stronger the association between experienced workplace bullying and negative effects on job satisfaction.

6.2 METHOD

6.2.1 Participants and Procedure

The study covered 1,051 teachers employed in 120 educational institutions. The age of the respondents ranged from 21 to 64 years. The mean age was 39.3 years (*SD* = 9.1). Women constituted a vast majority (86%). Almost all the respondents held a university degree (95%).

The study was conducted in September 2006 as a field study. The participation in the study was voluntary and anonymous. In total, 1,559 sets of questionnaires were distributed by interviewers, and 1,051 correctly completed, returned questionnaires (a fairly high response rate of 67.4%) were considered in the analysis.

6.2.2 Measures

6.2.2.1 NAQ questionnaire

The *Negative Acts Questionnaire* (NAQ) by Einarsen et al. (Einarsen et al. 1994; Einarsen and Raknes1997) in the Polish translation by Warszewska-Makuch (2007) was used to measure workplace bullying. The questionnaire consists of 22 questions reflecting various negative behaviors that may occur at work (not specifically referring to the *workplace bullying* term). Respondents are asked to indicate on a five-point scale (*never*, *less than once a month*, *once a month*, *once a week*, *daily*) the frequency of encounters with particular behaviors in the past six months. An additional question measuring the general feeling of being bullied at work, according to the presented definition of workplace bullying, was added to 22 items covering the negative experiences. Regarding the workplace bullying definition, respondents are asked to indicate whether they have been victims of bullying in the last six months and how often this has happened (*rarely*, *from time to time*, *several times a week*, *almost every day*). Hence, NAQ allows the measurement of both the experience of specific behaviors considered as bullying practices and the subjective feeling of being bullied. The Polish version of the NAQ scale is characterized by high reliability (*Cronbach's* α = 0.94). Three indicators of workplace bullying were used in the data analysis, i.e.:

1. Workplace bullying (global index) – achieved by aggregate score obtained by all respondents in 22 items of the NAQ questionnaire. This variable is of a quantitative nature. The higher the numerical score, the greater the intensity of experienced workplace bullying was considered.
2. Workplace bullying (number of negative actions) – this indicator refers to the number of negative actions experienced by the respondents. This variable is of a quantitative nature. The assumption was that the higher the numerical score, the more bullying practices the respondent experienced.

3. Workplace bullying (Leymann's index) – in turn, this index is based on the division of all respondents into two groups, i.e. victims of workplace bullying and *non-victims* of workplace bullying (control group), according to Leymann's criterion* (Leymann 1996). This variable is a two-tier variable where "1" stands for victims and "0" stands for *non-victims* of workplace bulling.

6.2.2.2 CISS Questionnaire

The Coping Inventory for Stressful Situations (CISS) developed by Endler and Parker (Endler 1994; 1990), adapted by Strelau, Jaworowska, Wrześniewski and Szczepaniak (Strelau et al. 2005), was used to measure the workplace bullying coping styles. The questionnaire consists of 48 statements, which make up three scales:

1. Task-oriented style (TOS) – defines a task-based style. High achievers on this scale take action to address the problem through cognitive transformation or change.
2. Emotion-oriented style (EOS) – is a style characterized by persons who, under pressure, concentrate on their own emotions, such as fear, anger, anxiety. It is also associated with a tendency to fantasizing and wishful thinking.
3. Avoidance-oriented style (AOS) has two subscales: engaging in substitution activities and seeking social contacts. This scale refers to a style based on avoiding the acknowledgment, analysis and experience of stressful situations (Strelau et al. 2005).

The respondent provides the answer on the five-point scale, according to the item's statement (48 items in total), selecting the option that best describes the frequency of the activity (1–5). In the Polish adaptation of the CISS, the reliability measured by the *Cronbach's* α index for the TOS and EOS scales ranged from 0.82 to 0.88, and for the AOS scales from 0.74 to 0.78 (Strelau et al. 2005).

6.2.2.3 GHQ-12 Questionnaire

The General Health Questionnaire (GHQ) (Goldberg 1978) was used to measure the mental health of adults. It mainly concerns two categories: inability to continue normal, *healthy* functioning, and the emergence of a new phenomenon, the *mental distress*. In the present study a shorter, 12-item version of GHQ was used. Each questionnaire item is a question asking whether the respondent has recently experienced a particular symptom or behaved in the way indicated in the question. The respondent marks his or her answer on a scale from *not at all* to *much more than usual*. The reliability index for the original version of the GHQ-12 was *Cronbach's* $\alpha = 0.85$ (Goldberg and Williams 2001).

---

* Assuming that a victim of workplace bullying is a person who has experienced at least one of the negative effects listed in the NAQ questionnaire. In addition, this experience of negative actions/practices in the workplace had to be repeated at least once a week and last at least six months.

6.2.2.4 Job Satisfaction Scale

The scale used to measure job satisfaction was derived from the Copenhagen Psychosocial Questionnaire (COPSOQ) by Kristensen (Hasselhorn 2003). The respondents were asked to respond to each item on a four-point scale indicating the degree of satisfaction with each aspect of the job (1 – *very dissatisfied*, 2 – *dissatisfied*, 3 – *satisfied*, 4 – *very satisfied*). The reliability of the Polish version scale in NEXT studies was *Cronbach's* $\alpha = 0.78$ (Hasselhorn 2003).

6.2.2.5 Demographic Variables

The demographic question set covers variables such as age, gender, marital status, educational attainment, occupation, employment sector, employment (full-time, part-time, etc.), employment contract (indefinite, fixed-term, etc.), employment status (based on employment relationship, self-employed, etc.), type of organization (private, public, etc.), number of employees in the organization and position held in the organization.

6.2.2.6 Control Variables

Variables, i.e. psychosocial working conditions and impactful life events were controlled by analyzing the associations between workplace bullying, health, job satisfaction and stress-coping styles. The individual tools with which these variables were examined are described below.

6.2.2.6.1 Psychosocial Working Conditions

Variables, i.e. demands, control, development opportunities and social support at work, were assessed with the COPSOQ scales, developed by Kristensen (Hasselhorn 2003). The Polish version was developed by Widerszal-Bazyl and Radkiewicz within the NEXT project (Hasselhorn 2003). The reliability of the Polish version of these scales ranged from *Cronbach's* $\alpha = 0.72$ to *Cronbach's* $\alpha = 0.83$ (Polish sample N=3106) (Hasselhorn 2003).

6.2.2.6.2 Impactful Life Events

In addition, the questionnaire set was expanded by two questions to check whether the respondents had experienced particularly painful and/or particularly joyful life events in the last six months.

6.2.3 Statistical Analysis

Data analyses were performed using the SPSS 11.5 and AMOS 5.0 software. First, descriptive statistics were calculated for each variable (internal consistency, means, standard deviations, minimum and maximum values, skewness, kurtosis and normal distribution tests). Then, to assess the strength and significance of the relationships between these variables, *r-Pearson* correlation analyses were carried out. In the last stage, a number of hierarchical regression analyses (cf. Krejtz and Krejtz 2007), and an interactional effects* regression analysis (cf. Bedyńska and Książek 2012)

* In order to calculate the interactional component, the independent variables were centralized.

were carried out. These analyses verified whether particular styles of coping with stress were significant moderators in the relationship between workplace bullying and employee well-being. The hierarchical regression structure controlled the effect of selected demographic variables and psychosocial factors of the working environment on dependent variables.

6.3 RESULTS

6.3.1 Descriptive Statistics

The results of the internal consistency (measured by *Cronbach's α* index), means, standard deviations, minimum and maximum values, skewness, kurtosis and *Kolmogrow–Smirnow* normal distribution tests of the measured variables are presented in Table 6.1.

As mentioned earlier, the workplace bullying variable was measured by three indicators, i.e. the global index, the number of negative actions and the Leymann's index. The descriptive statistics of the first two indexes as quantitative variables are presented in Table 6.1. The third, i.e. the Leymann's index, is a two-tier variable ("0" – *non-victims of workplace bullying*, "1" – *victims of workplace bullying*), which allowed the identification of two subgroups corresponding to these

TABLE 6.1
The Study Sample Descriptive Statistics and Normal Distribution Tests of the Measured Variables

Scale	Variable	*N*	*α*	*M*	*SD*	*Min.*	*Max.*	*Skew.*	*Kurtosis*	*Z*	*p*
NAQ	Workplace bullying (global index)	1,048	0.94	28.55	8.74	22	92	2.76	11.31	7.34	0.001
	Workplace bullying (number of negative actions)	1,048	–	0.30	1.47	0	18	7.79	71.82	15.78	0.001
COPSOQ	Job demands	1,037	0.77	9.62	2.57	3	15	0.15	–0.59	3.64	0.001
	Social support	1,022	0.77	13.43	3.78	4	44	0.30	3.34	2.51	0.001
	Job control	1,021	0.86	13.81	3.69	4	42	–0.21	–0.48	2.58	0.001
	Personal development	1,044	0.51	8.30	1.39	2	10	–0.98	1.36	6.38	0.001
CISS	Task-oriented style	993	0.82	58.36	8.19	31	105	0.29	1.43	1.13	0.156
	Emotion-oriented style	983	0.84	43.25	10.08	17	89	0.23	0.28	1.23	0.098
	Avoidance-oriented style	978	0.71	44.02	7.88	19	88	0.23	1.32	1.18	0.123
GHQ-12	Mental health	1,033	0.90	12.19	5.84	0	36	1.09	1.62	4.85	0.001
COPSOQ	Job satisfaction	1,010	0.81	15.94	2.89	6	26	–0.22	0.53	3.92	0.001

α – *Cronbach's α* internal consistency index.
Z – normal distribution test (*Kolmogrow–Smirnov* test).
$p < 0.05$ – sample distribution (does not follow normal distribution).

TABLE 6.2
Classification of the Respondents as Victims and Non-Victims of Workplace Bullying (N = 1,051)

Group	Frequency	Percentage
Non-victims (control group)	958	91
Victims	93	9
Total	1,051	100

categories in the entire study sample. Table 6.2 shows the numerical and percentage distribution of the respondents within this variable. Ninety-three victims of workplace bullying were identified (9% of all respondents) in the entire study group.

The correlation analysis results are presented in Table 6.3.

Gender difference significance test results of the studied variables are presented in Table 6.4.

6.3.2 Regression Analysis

In order to verify the study hypotheses, regression analyses were carried out, separately for each stress-coping style and for each of the two dependent variables, i.e. mental health and job satisfaction. In steps 1, 2 and 3 of the regression analysis, control variables were introduced, i.e. age, gender, impactful life events and psychosocial working conditions. Step 4 included a specific coping style (task-oriented style, emotion-oriented style or avoidance-oriented style, respectively). In step 5, the workplace bullying variable was introduced as the Leymann's index. In step 6 of the analysis, an interaction factor was introduced, i.e. workplace bullying (Leymann's index) × task-oriented style. The workplace bullying i.e. the Leymann's index choice, was dictated by preliminary analysis results which showed that the index was the only factor out of all three workplace bullying indexes to play a significant role in interacting with one of the coping styles, i.e. the task-oriented style, for both mental health prediction and job satisfaction. Regarding the latter variable, the workplace bullying global index, although insignificant to mental health, was also found to be an important indicator of job satisfaction, and to interact with the task-oriented style.

6.3.2.1 Workplace Bullying – Stress-Coping Styles' Interaction and Mental Health Status

Table 6.5 presents the mental health and the workplace bullying–task-oriented style interaction component hierarchical regression analysis results.

The task-oriented style (TOS) was introduced to the model in step 4 of the analysis. It was found that TOS was an important predictor of mental health ($\beta = -0.21$; $p < 0.001$), explaining 4% of the variable variance. The negative beta coefficient

TABLE 6.3
Correlation Matrix (*r-Pearson*) of the Study Variables

No.	Variable	1	2	3	4	5	6	7	8	9	10	11	12
1	Workplace bullying (global index)	–											
2	Workplace bullying (number of negative actions)	0.73**	–										
3	Workplace bullying (Leymann's index)	0.57**	0.58**	–									
4	Task-oriented style	–0.07*	0.03	0.01	–								
5	Emotion-oriented style	0.17**	0.07*	0.05	–0.28**	–							
6	Avoidance-oriented style	0.03	0.00	0.01	0.04	0.22**	–						
7	Mental health	0.31**	0.17**	0.13**	–0.26**	0.45**	0.01	–					
8	Job satisfaction	–0.46**	–0.27**	–0.26**	0.14**	–0.17**	0.04	–0.36*	–				
9	Job demands	–0.39**	–0.21**	–0.21**	0.02	–0.14**	–0.03	–0.29**	0.36**	–			
10	Social support	–0.45**	–0.20**	–0.22**	0.12**	–0.14**	0.12**	–0.26**	0.51**	0.29**	–		
11	Job control	–0.42**	–0.25**	–0.24**	0.23**	–0.23**	0.04	–0.29**	0.52**	0.35**	0.47**	–	
12	Development opportunities	–0.20**	–0.11**	–0.09**	0.25**	–0.06	0.08*	–0.13**	0.39**	0.01	0.28**	0.38**	–
13	Age	0.02	0.01	–0.04	0.10**	–0.01	–0.07*	0.05	0.04	–0.09**	–0.11	0.16**	0.10**

* Correlation is significant at $p < 0.05$ (bivariate test).
**Correlation is significant at the level of $p < 0.01$ (bivariate test).

TABLE 6.4
Comparison of Men and Women in Terms of Experienced Workplace Bullying (Global Index and the Number of Negative Actions), Mental Health Status, Job Satisfaction, Stress-Coping Styles and Psychosocial Working Conditions: Student's *t* and *Cohen's d* Difference Significance Test Results

	Men		Women		t	
Variable	*M*	*SD*	*M*	*SD*	*t*	*Cohen's d*
Workplace bullying (global index)	29.27	9.04	28.33	8.46	1.21	0.08
Workplace bullying (number of activities)	0.35	1.33	0.27	1.39	0.63	0.04
Mental health	10.93	5.17	12.38	5.93	–2.68**	0.17
Job satisfaction	16.50	3.28	15.87	2.82	2.38*	0.15
Task-oriented style	58.17	8.47	58.42	8.15	–0.34	0.02
Emotion-oriented style	40.34	10.52	43.69	9.96	–3.55***	0.23
Avoidance-oriented style	42.35	8.26	44.29	7.81	–2.64**	0.17
Job demands	10.20	2.59	9.53	2.55	2.83**	0.18
Social support	13.58	3.83	13.43	3.77	0.45	0.03
Job control	13.89	3.47	13.82	3.70	0.22	0.01
Personal development	8.14	1.61	8.33	1.35	–1.44	0.09

*** $p < 0.001$; ** $p < 0.01$; * $p < 0.05$.

TABLE 6.5
Results of the Hierarchical Regression Analysis of Mental Health in Relation to TOS, Workplace Bullying (Leymann's Index) and the TOS–Workplace Bullying Interaction (Controlled by Age, Gender, Experience of Painful and Joyful Events, Job Demands, Social Support, Job Control and Development Opportunities) in the Study Sample (*N* = 883)

Step and Predictor	R^2	ΔR^2	*F*	β
Step 4	0.30	0.04***	40.96***	
Task-oriented style				–0.21***
Step 5	0.30	0.002	37.10***	
Task-oriented style				–0.21***
Workplace bullying (Leymann's index)				0.05
Step 6	0.31	0.01***	34.71***	
Task-oriented style				–0.24***
Workplace bullying (Leymann's index)				0.04
Task-oriented style × workplace bullying (Leymann's style)				0,09**

*** $p < 0.001$; ** $p < 0.01$; * $p < 0.05$.

indicates that the higher the level of TOS, the better the mental health. In turn, the workplace bullying introduced in step 5 of the regression analysis proved to be an insignificant predictor of mental health ($\beta = 0.05$; $p > 0.05$). In the sixth, last step of the regression analysis, an interaction of task-oriented style and workplace bullying was introduced, and was found to be an important predictor of mental health ($\beta = 0.09$; $p < 0.01$), explaining 1% of the variable variance. The model fit the data in the last step of the regression analysis $F(11.871) = 34.71$; $p < 0.001$ and explained 31% of the mental health variance. The direction of this interaction is further discussed below, accounting for additional analyses.

The analysis of the relationship between task-oriented style and mental health in the groups identified on the basis of workplace bullying experiences, i.e. victims and *non-victims* of workplace bullying, demonstrated that this relationship was insignificant in the victims group ($\beta = -0.03$; $p > 0.05$) (Table 6.5).

In the *non-victims* group, however, the relationship between the task-oriented style and mental health proved to be significant ($\beta = -0.24$; $p < 0.001$). The negative beta coefficient indicates that in the control group (*non-victims* of workplace bullying) a high level of task-oriented style was associated with a low level of mental health problems. The model fit the data $F(9.795) = 37.12$; $p < 0.001$ and explained 30% of the mental health variance (Table 6.6).

The earlier results are illustrated in Figure 6.2 showing the effect of the interaction of task-oriented style and workplace bullying on mental health. The victims of workplace bullying have a constant level of mental health status regardless of the frequency of the task-oriented style used in coping with stress. The *non-victims* of workplace bullying suffer from a larger number of mental health problems when they rarely use the task-oriented style.

In order to verify whether there are significant differences in the mental health status, depending on the level of task-oriented coping style, among both the victims and *non-victims* of workplace bullying, a one-way variance analysis was carried out, i.e. separately for the group of victims and the group of *non-victims* of workplace bullying. A statistically significant task-oriented style effect was obtained in the group of *non-victims* $F(2.893) = 37.12$; $p < 0.001$ (Table 6.7). Table 6.8 shows that *post hoc* Games–Howell test comparisons revealed significant differences in all three compared group categories. The results demonstrated

TABLE 6.6

Results of the Hierarchical Regression Analysis of Mental Health in Relation to TOS (Controlled by Age, Gender, Experience of Painful and Joyful Events, Job Demands, Social Support, Job Control and Development Opportunities) in the Group of Workplace Bullying Victims ($N = 78$)

Step and Predictor	R^2	ΔR^2	F	β
Step 4	0.28	0.001	2.88**	
Task-oriented style				–0.03

*** $p < 0.001$; ** $p < 0.01$; * $p < 0.05$.

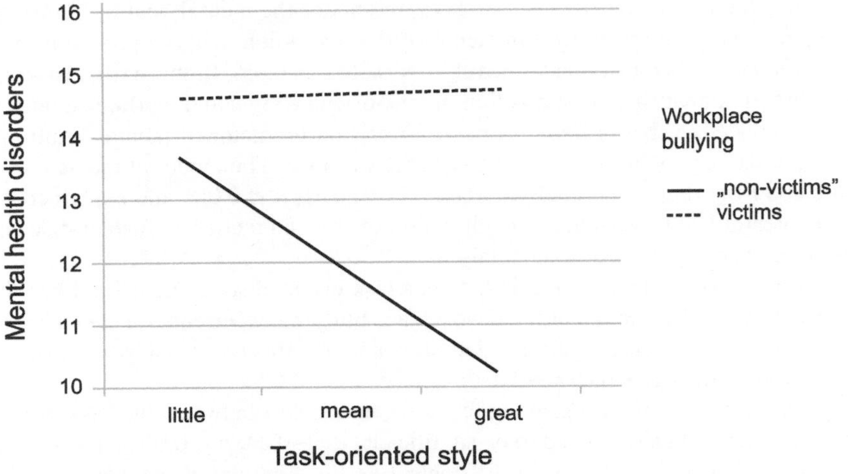

FIGURE 6.2 Victim of workplace bullying and the task-oriented style interaction effects on mental health status.

TABLE 6.7
Results of the Hierarchical Regression Analysis of Mental Health in Relation to TOS (Controlled by Age, Gender, Experience of Painful and Joyful Events, Job Demands, Social Support, Job Control and Development Opportunities) in the Group of *Non-Victims* of Workplace Bullying (N = 805)

Step and Predictor	R^2	ΔR^2	F	β
Step 4	0.30	0.05***	37.12***	
Task-oriented style				–0.24***

*** $p < 0.001$; ** $p < 0.01$; * $p < 0.05$.

TABLE 6.8
Mental Health of Low-, Medium- and High-Level of Task-Oriented Style *Non-Victims* Group

	Level of Task-Oriented Style in Coping with Workplace Bullying									
	Low (1)*			Medium (2)			High (3)			
	N	M	SD	N	M	SD	N	M	SD	Post hoc
Mental health disorders	285	14.06	6.25	301	11.95	4.87	310	10.08	5.32	1> 21> 32> 3

* The numbers in the column headings correspond to those used in the study to show significant differences in the *Post hoc* column.

that the group with a low level of task-oriented style differed significantly from the group with a medium level of task-oriented style, and also from the group with a high level of task-oriented style. Moreover, the group with a medium level of task-oriented style differed significantly from the group with a high level of task-oriented style .

Analogous analyses carried out in the *victims* group showed that the task-oriented style of coping with stress had no statistically significant effect on mental health, $F(2.81) = 1.44$; $p > 0.05$.

The analysis demonstrated that in the case of the remaining two styles of coping with stress, their moderating role in the relationship between workplace bullying and mental health was not significant.

6.3.2.2 Workplace Bullying – Stress-Coping Styles' Interaction and Job Satisfaction

Detailed regression analysis results, testing the significance of the interaction of the task-oriented style with the experience of workplace bullying as a predictor of job satisfaction, are presented in Table 6.9. The control variables explained as much as 45% of the job satisfaction variance (only the age was insignificant). In turn, the task-oriented style introduced in step 4 of the analysis did not prove an important predictor of the dependent variable. The Leymann's index-measured workplace bullying, introduced in step 5 of the regression analysis, although significant ($\beta = -0.07$; $p < 0.01$), explained only 0.4% of the job satisfaction variance (<1%). Similarly, the interaction of the task-oriented style with workplace bullying introduced in step 6 of the analysis was found to be significant ($\beta = -0.07$; $p < 0.05$); however, it explained

TABLE 6.9
Results of the Stepwise Regression Analysis of Job Satisfaction in Relation to TOS, Workplace Bullying (Leymann's Index) and the TOS–Workplace Bullying Interaction (Controlled by: Age, Gender, Experiencing Painful and Joyful Events, Job Demands, Social Support, Job Control and Development Opportunities) ($N = 858$)

Step and Predictor	R^2	ΔR^2	F	β
Step 4	0.45	0.000	76.87***	
Task-oriented style				0.01
Step 5	0.45	0.004*	70.17***	
Task-oriented style				0.01
Workplace bullying (Leymann's index)				–0.07*
Step 6	0.46	0.01*	64.73***	
Task-oriented style				0.03
Workplace bullying (Leymann's index)				–0.07**
Task-oriented style × workplace bullying (Leymann's index)				–0.07*

*** $p < 0.001$; ** $p < 0.01$; * $p < 0.05$.

only 1% of the job satisfaction variance. The last step of the analysis model had an adjusted data fit $F(11.846) = 64.73$; $p < 0.001$, and its predictors explained 46% of the dependent variable variance.

The analysis of the relationship between the task-oriented style and job satisfaction in the groups identified on the basis of workplace bullying experiences, i.e. victims and *non-victims* of workplace bullying, showed that in the victim group the relationship was insignificant ($\beta = -0.16$; $p > 0.05$) (Table 6.10). Conversely, in the *non-victims* group, the relationship between the task-oriented style and job satisfaction was significant ($\beta = 0.17$; $p < 0.001$). The positive beta coefficient indicates that in the control group (*non-victims* of workplace bullying) a high level of task-oriented style was associated with a high level of job satisfaction. This model had a good data fit $F(9.779) = 71.10$; $p < 0.001$ and explained 44% of the job satisfaction variance (Table 6.11).

The above analyses are illustrated in Figure 6.3 showing the effects of the interaction between the task-oriented style and workplace bullying on job satisfaction. It reveals that the more the task-oriented style was used in stressful situations by persons in the *non-victims* group, the greater their job satisfaction was. In turn, the high level of task-oriented style was associated with lower job satisfaction among the victims of workplace bullying. Although, as the abovementioned means comparison has shown, these differences are not statistically significant.

TABLE 6.10
Results of the Hierarchical Regression Analysis of Job Satisfaction in Relation to TOS (Controlled by: Age, Gender, Experience of Painful and Joyful Events, Job Demands, Social Support, Job Control and Development Opportunities) in the Group of Workplace Bullying Victims ($N = 76$)

The step and the Predictor	R^2	ΔR^2	F	β
Step 4	0.51	0.01	7.74***	
Task-oriented style				–0.16

*** $p < 0.001$; ** $p < 0.01$; * $p < 0.05$.

TABLE 6.11
Results of the Hierarchical Regression Analysis of Job Satisfaction in Relation to TOS (Controlled by: Age, Gender, Experience of Painful and Joyful Events, Job Demands, Social Support, Job Control and Development Opportunities) in the Group of *Non-Victims* of Workplace Bullying ($N = 789$)

Predictor and Step	R^2	ΔR^2	F	β
Step 4	0.44	0.03***	71.10***	
Task-oriented style				0.17***

*** $p < 0.001$; ** $p < 0.01$; * $p < 0.05$.

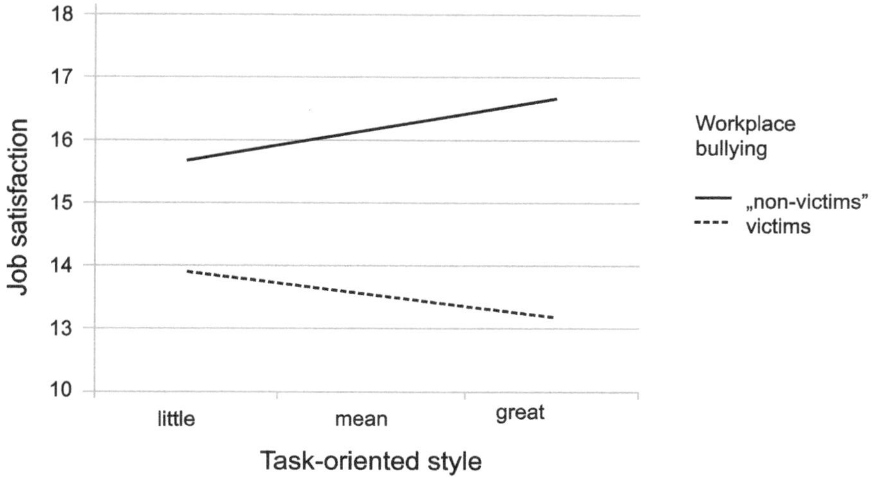

FIGURE 6.3 Victim of workplace bullying and the task-oriented style interaction effects on job satisfaction.

In order to verify whether there were significant differences in job satisfaction levels among both victims and *non-victims* of workplace bullying in relation to the task-oriented style used in coping with stress, a one-way variance analysis was carried out, i.e. separately for the two identified groups of victims and *non-victims* of workplace bullying. A statistically significant effect of the task-oriented style was observed in the group of *non-victims* $F(2.873) = 11.27; p < 0.001$. However, as shown in Table 6.12, the Scheffe's *post hoc* test comparisons revealed that significant differences did not apply to all groups. It was found that the group with a low level of task-oriented style differed significantly from the group with a medium level of task-oriented style, and also from the group with a high level of task-oriented style, whereas the group with a medium level of task-oriented style did not differ significantly from the group with a high level of task-oriented style.

TABLE 6.12
Job Satisfaction of Low-, Medium- and High-Level Task-Oriented Style Non-Victims Groups

	Level of Task-Oriented Style in Coping with Workplace Bullying									
	Low (1)*			Medium (2)			High (3)			
	N	*M*	*SD*	*N*	*M*	*SD*	*N*	*M*	*SD*	*Post hoc*
Job satisfaction	277	15.62	2.80	294	16.24	2.48	305	16.68	2.84	1 < 2. 1< 3. 2 = 3

* The numbers in the column headings correspond to those used in the study to show significant differences in the *Post hoc* column.

Analogous analysis conducted in the victims group revealed that the task-oriented style had no statistically significant effect on job satisfaction, $F(2.77) = 0.07$; $p > 0.05$.

The role of the interaction between the emotion-oriented style and workplace bulling as a predictor of job satisfaction was not significant. Also, the stress coping style itself was not a significant predictor of job satisfaction.

Similarly, the results concerning the relationship between the interaction of workplace bullying with the avoidance-oriented style, and the level of job satisfaction showed that this interaction was not a significant predictor of the dependent variable. The avoidance-oriented style itself was also not a significant predictor of job satisfaction either.

6.4 DISCUSSION OF RESULTS

In line with the hypotheses, it was expected that stress management styles would moderate the relationship between experienced workplace bullying and employee well-being. Equally, the premise indicated that the direction of this moderation would be such that the task-oriented style would mitigate the relationship between workplace bullying and poorer well-being (namely, poorer mental health and lower job satisfaction), while the emotion- and avoidance-oriented styles would exacerbate this relationship.

The obtained results did not confirm the hypothesis put forward. It has not been found that any of the stress-coping styles analyzed, be it task-oriented style, emotion-oriented style or avoidance-oriented style, have any impact on the relationship between workplace bullying and negative effects on mental health and job satisfaction.

6.4.1 Task-Oriented Style and the Relationship between Experienced Workplace Bullying, Mental Health and Job Satisfaction

The task-oriented style hypotheses have been formulated according to the ability to cope with stress in general. Although the effectiveness of this style is not entirely clear (e.g. Heszen-Niejodek 1997; Heszen-Niejodek and Sęk 2007), there have been reports indicating that more often than the other styles, it prevents negative effects of stress on health (see Ogińska-Bulik and Juczyński 2008). However, the present study results suggest that in terms of workplace bullying, this type of stress management style has not proven effective. This may be due to the limited capacity of ruling out further acts of workplace bullying. Hence, the task-oriented style does not prevent the prolonged negative impact of workplace bullying on worker well-being. The literature does not offer any accounts of a study that would directly report on the task-oriented style as approached in the current study, i.e. a general inclination to undertake a problem-solving action (and tested by statements such as: *I take immediate and appropriate action*; *I strive to control the situation*; *I try to organize things in such a way as to control the situation*). Nevertheless, there have been studies that have accounted for specific behavioral patterns of coping with workplace bullying, which can be seen as manifestations of the task-oriented style. These studies have supported the aforementioned interpretation of the results as evidence proving the ineffectiveness of the task-oriented style in coping with workplace bullying.

The task-oriented style, as a response to the workplace bullying stressor, may entail aggressive behaviors. Disruptive actions are a kind of retaliation against the perpetrator, an attempt to change his/her behavior by punishing the bully or manifesting the victim's readiness to defend himself/herself against further ill-treatment. Many consider such aggression as a morally justified response to harassment (Aquino and Thau 2009; Zapf and Gross 2001). Ostracism, which is often related to workplace bullying, threatens fundamental human needs, including the need for control and recognition. This, according to Williams (2007), is the reason for the victims' aggression (cf. also Leary et al. 2006). In a study conducted by Zellars et al. (Zellars et al. 2002) the researchers equally revealed that workers who felt hostility from the supervisor presented less civic organizational behaviors and even perceived such behaviors as an *extra* effort (going beyond their organizational duties). Researchers have treated this phenomenon as a kind of retaliation against the manager who victimizes the employee. Lee and Brotheridge (Lee and Brotheridge 2006) showed that workers who had been bullied at work were at the same time more inclined to undermine co-workers. In the light of these studies, the task-oriented style of coping with stress in relation to workplace bullying may often take the form of aggressive behaviors.

Moreover, the use of a confrontational, aggressive, task-oriented style by the victim may exacerbate the harassing behavior by the bully. The hostile relationship between the perpetrator and the victim may then escalate, leading to an endless spiral of aggression (Aquino and Thau 2009). Such a depiction has also been supported by Williams' (2007) observations, who has emphasized that that ostracism, exclusion and humiliation of individuals may often be associated with a decline in social behavior and a simultaneous emergence of an anti-social attitude toward others, which in turn may lead to even greater social exclusion.

It therefore seems that an aggressive, task-oriented style is ineffective in coping with workplace bullying and therefore does not improve well-being either. However, it is worth noting that the task-oriented style not only involves aggressive behavior, but may also entail a friendly attitude toward others, including the bully himself/herself. Researchers have reported on cases of workplace bullying victims who adopted a *tend-and-befriend* strategy (Williams 2007) and despite the experienced ostracism tried to maintain a friendly demeanor (Gardner et al. 2000; Pickett and Gardner 2005). However, there have been accounts of such methods as ineffective ways of curbing down the workplace bullying. Cortina and Magley (Cortina 2003) revealed that both employees who confronted the perpetrator (aggressive style) and those who sought support from other co-workers (non-aggressive style) met with a negative reaction from the social environment. The difference between the groups consisted in the type of aggression they experienced as a result of the undertaken strategy: the former – mainly aggression concerning the work itself (forced transfer, dismissal) – and the latter – aggression aimed at social relations (ostracism, intimidation, harassment). Similar observations were made by Williams (2007), who has stressed that many of the pro-social behaviors are not effective and do not produce results that meet the expectations of the individual who shows them. Most times, attempts to be more socially acceptable lead to even stronger attacks on the victim who becomes an easy target for social manipulation. Zapf and Gross (Zapf and Gross 2001) have also noted that in the case of workplace bullying, the use of a non-aggressive cooperative strategy, which is usually recommended as the most effective strategy for solving long-term conflicts (Thomas 1992), has proven ineffective.

In summary, the most feasible understanding of the present study outcomes, which do not support the claimed task-oriented style's capacity to mitigate the negative relationship between workplace bullying and well-being, is that the task-oriented style does not prevent workplace bullying and, consequently, cannot inhibit its negative effects on mental health or job satisfaction.

This is particularly striking in the light of another result obtained in this study, which has shown that in the non-victims group the task-oriented style was associated with an improved well-being (better mental health and greater job satisfaction). It would be rather naïve to believe that a large group of *non-victims* had experienced no forms of stress at work. Certain forms of stress may have been present and may have led to poorer well-being. However, in this subgroup the task-oriented stress-coping style was conducive to improved well-being. This proves that the task-oriented style of coping with stress – and this has been the underlying hypothesis – is effective in relation to most stressors in the workplace. Workplace bullying, however, is such a specific and extreme kind of stressor, that no stress-coping style is effective in coping with it.

6.4.2 Emotion- and Avoidance-Oriented Style and the Relationship between Experienced Workplace Bullying, Mental Health and Job Satisfaction

Similar conclusions can also be drawn from the analyzed roles of the remaining two stress management styles, i.e. emotion- and avoidance-oriented styles. The hypothesis put forward has not been confirmed: the use of the two stress management styles has not exacerbated the negative effects of workplace bullying on employee well-being.

The present study results may indicate that these styles do not play such a negative role in the experiences of workplace bullying as has been assumed. In the current study the emotion-oriented style has been referred to as focusing on negative emotions (*I am worried that I will not manage, I'm blaming myself for caring about it too much*), and also a state of bewilderment (*I'm feeling perplexed and do not know what to do*). Perhaps the lack of response to workplace bullying by virtue of such a perplexity reduces the likelihood of a dysfunctional reaction (aggression against the perpetrator), and consequently does not further escalate the bullying behaviors, inhibiting the negative impact of workplace bullying on health. Williams (2007) has compared such a behavioral pattern to the animal world, where in certain situations a fighting or fleeing reaction can be ineffective or even dangerous. By analogy, the retaliation (task-oriented style) becomes irrational also for persons experiencing ostracism in the working environment, as it may further exacerbate the exclusion of the individual from the work group.

The avoidance-oriented style explored in the study has been identified as a tendency to avoid acknowledging and analyzing stressful experiences (Strelau et al. 2005). This style has taken two forms in the study, i.e. engaging in substitute activities (*"I try to fall asleep"*, *"I eat my favorite dish"*), and seeking social contacts (*I spend time with a close person, I visit a friend*). The avoidance-oriented style behaviors rather seem to avoid a confrontation with the perpetrator. Such an evasive attitude may not actually be associated with negative results. This is in line with the findings of Aquino and Thau (Aquino 2009), who have observed that in certain situations the avoidance-oriented style may even be the most effective coping strategy, since, similarly to the *bewilderment*, it reduces the frequency of bullying incidents and prevents a further escalation of the conflict, thus minimizing the costs for the victim.

6.5 STRENGTHS AND LIMITATIONS

A major limitation of the present research has been the cross-sectional character of the study. Consequently, the relationships between variables tested cannot be defined in terms of causal associations. A further drawback is related to the selection of the study sample. The survey covered only one occupational group, and hence the results cannot be extended to the entire population of Polish workers.

Furthermore, the sheer structure of the workplace bullying victims group identified in the study has also been equally important. These were teachers who continued to work full-time. Therefore, the individuals most affected by workplace abuse may have actually not taken part in the study. Consequently, the analyzed results may correspond mainly to persons who had experienced workplace bullying at a medium or low level. Such a study sample composition may have thus underplayed the associations between workplace bullying and the other analyzed factors. It is also worth noting that most studies in the field have focused on those workplace bullying victims who were no longer professionally active as a result of poor health. As the present study has reported on the victims of workplace bullying who were still in employment, the comparison of the study results with other research findings has been problematic.

Moreover, the present study has accounted for the self-reported data which may constitute an additional weakness, since the study may have been subject to the common-method bias error, overestimating the identified correlations.

Some difficulties may also relate to the stress-coping styles defined in the study. The coping strategies have been perceived as individual response skills, according to the three broadly defined stress-coping styles. Future research tools should thus account for the variety of coping strategies determined by the specificity of workplace bullying experience and its phase. It seems particularly important to consider the diversity of reactions corresponding to the task-oriented style as they may entail both aggressive and friendly behaviors, yielding different results.

Regarding the emotion-oriented style, it would be interesting to introduce an additional, so far omitted factor, i.e. the effect of forgiving the perpetrator for the suffering caused, in future studies. This strategy has been defined in the literature as an effort made by the victim to transform negative emotions and thoughts about the perpetrator into neutral or even positive feelings (Aquino et al. 2006). However, despite the existing evidence supporting that the willingness to forgive can neutralize the negative consequences of being bullied (cf. Freedman and Enright 1996), no empirical studies have been conducted so far to verify the effects of forgiving in the workplace bullying context.

The strength of the present research has been the very account of workplace bullying, as there have been very few studies in the field carried out in Poland to this day. This has made it possible to verify some dependencies so far identified only in research conducted in other countries.

A further advantage of the present research has been the focus on several workplace bullying indicators. The three indices of the workplace bullying variable have allowed in-depth analyses and proved their relevance to future studies.

6.6 CONCLUSIONS AND PRACTICAL IMPLICATIONS

In summary, the obtained results may prove that all stress-coping styles are insignificant when coping with workplace bullying. They neither restrain (as expected from the task-oriented style) nor exacerbate the harassing behavior (as expected from the emotion- and avoidance-oriented styles) This seems to stem from the nature of workplace bullying, which is the trauma associated with the perpetuated, specific type of violence. Therefore, workplace bullying can be considered an objective, universal stressor that affects all individuals, regardless of the subjective perception of the bullying experience. Several studies (cf. Hogh et al. 2011) have shown that, despite the various efforts made by victims to stop the violent practice, the majority find it difficult to do so without third-party support (Hogh and Dofradottir 2001; Zapf and Gross 2001). This stems from the low predictability and controllability of the workplace bullying experience, whereby victims feel perplexed, guilty, and depressed and develop low self-esteem. As Hogh, Mikkelsen and Hansen (Hogh et al. 2011) have emphasized, victims have little, if any, external and internal resources to cope with workplace bullying and, worse still, are painfully aware of this. Referring to Hobfoll's Conservation of Resources Theory (COR) (1989), it can be concluded that, precisely because of the lack of adequate resources, victims of workplace bullying cannot resist it. This lack of resources and the associated sense of bewilderment may be mainly due to the insufficient number of measures taken to support the victims in struggling with workplace bullying, such as lack of established organizational anti-harassment policies, hence leaving the victims perplexed and alone.

In this view, the best strategy to respond to violence experienced at work seems to be quitting the job. Such conclusions have also been proposed by German researchers (Zapf and Gross 2001), who have concluded that when faced with workplace bullying no coping strategies are effective – apart from changing jobs and a definitive separation from the perpetrator. Such a sad conclusion has important practical implications. Restricting workplace bullying practices not only should focus on supporting the victims in coping with the problem, but equally on encouraging the organization to address workplace bullying and, ideally, prevent it.

An important value of the present research is its practical implications. Thus, the results indicating the ineffectiveness – in the face of workplace bullying – of individual stress-coping styles suggest that the most appropriate form of reducing the negative effects of workplace bullying is its prevention at organizational level, i.e. manager training, implementation of anti-harassment and workplace bullying policies in the organization and education of workers in this regard.

REFERENCES

Antonovsky, A. 1995. *Rozwikłanie tajemnicy zdrowia. Jak radzić sobie ze stresem i nie zachorować.* Warszawa: Instytut Psychiatrii i Neurologii.

Aquino, K., and S. Thau. 2009. Workplace victimization: Aggression from the target's perspective. *Annu Rev Psychol* 60:717–741.

Aquino, K., T. M. Tripp, and R. J. Bies. 2006. Getting even or moving on? Power, procedural justice and types of offense as predictors of revenge, forgiveness, and avoidance in organizations. *J Appl Psychol* 91(3):653–668.

Bedyńska, S., and M. Książek. 2012. *Statystyczny Drogowskaz 3. Praktyczny przewodnik wykorzystania modeli regresji oraz równań strukturalnych.* Warszawa: Wydawnictwo Akademickie Sedno.

Berkowitz, L. 1989. Frustration-aggression hypothesis: Examination and reformulation. *Psychol Bull* 106(1):59–73.

Bowling, N. A., and T. A. Beehr. 2006. Workplace harassment from victim's perspective: A theoretical model and meta-analysis. *J Appl Psychol* 91(5):998–1012.

Cooper, C. L., and J. Marshall. 1976. Occupational sources of stress: A review of the literature relating to coronary heart disease and mental ill health. *J Occup Psychol* 49(1):11–28. DOI: 10.1111/j.2044-8325.1976.tb00325.x.

Cortina, L. M., and V. J. Magley. 2003. Raising voice, risking retaliation: Events following interpersonal mistreatment in the workplace. *J Occup Health Psychol* 8(4):247–265.

Djurkovic, N., D. McCormack, and G. Casimir. 2004. The physical and psychological effects of workplace bullying and their relationship to intention to leave: A test of the psychosomatic and disability hypotheses. *Int J Organ Theor Behav* 7(4):469–497. DOI: 10.1108/IJOTB-07-04-2004-B001.

Einarsen, S. 1999. The nature and causes of bullying at work. *Int J Manpow* 20(1–2):16–27.

Einarsen, S. 2000. Harassment and bullying at work: A review of the Scandinavian approach. *Aggress Violent Behav* 5(4):379–401.

Einarsen, S., H. Hoel, and G. Notelaers. 2009. Measuring exposure to bullying and harassment at work: Validity, factor structure and psychometric properties of the Negative Acts Questionnaire-Revised. *Work Stress* 23(1):24–44. DOI: 10.1080/02678370902815673.

Einarsen, S., H. Hoel, D. Zapf, and C. L.Cooper. 2003. The concept of bullying at work: European tradition. In: *Bullying and Emotional Abuse in the Workplace: International Perspectives in Research and Practice*, eds. S. Einarsen, H. Hoel, D. Zapf, and C. L. Cooper, 3–30. London: Taylor & Francis.

Einarsen, S., S. B. Matthiesen, and A. Skogstad. 1998. Bullying, burnout and well-being among assistant nurses. *J Occup Health Saf* 14:563–568.

Einarsen, S., and B. I. Raknes. 1997. Harassment in the workplace and the victimization of men. *Violence Vict* 12(3):247–263. DOI: 10.1891/0886-6708.12.3.247.

Einarsen, S., B. I. Raknes, and S. B. Matthiesen. 1994. Bullying and harassment at work and their relationships to work enviroment quality: An exploratory study. *Eur J Work Organ Psychol* 4:381–401.

Endler, N. S., and J. D. A. Parker. 1990. Multidimensional assessment of coping: A critical theory evaluation. *J Pers Soc Psychol* 58(5):844–854. DOI: 10.1037/0022-3514.58.5.844.

Endler, N. S., and J. D. A. Parker. 1994. Assessment of multidimensional coping: Task, emotion, and avoidance strategies. *Psychol Assess* 6(1):50–60. DOI: 10.1037/1040-3590.6.1.50.
Fraser, D. F., and J. F. Gilliam. 1987. Feeding under predation hazard: Response of the gruppy and Hart's rivulus from sites with contrasting predation hazard. *Behav Ecol Sociobiol* 21(4):203–209.
Freedman, S. R., and R. D. Enright. 1996. Forgiveness as an intervention goal with incest survivors. *J Consult Clin Psychol* 64(5):983–992.
Gardner, W. L., C. L. Pickett, and M. B. Brewer. 2000. Social Exclusion and Selective Memory: How the Need to belong Influences Memory for Social Events. *Pers Soc Psychol Bull* 26(4):486–496. DOI: 10.1177/0146167200266007.
Glasø, L., E. Bele, M. B. Nielsen, and S. Einarsen. 2011. Bus drivers' exposure to bullying at work: An occupation-specific approach. *Scand J Psychol* 52(5):484–493. DOI: 10.1111/j.1467-9450.2011.00895.x.
Goldberg, D. 1978. *Manual of the General Health Questionnaire*. Windsor: NFER-NELSON.
Goldberg, D., and P. Williams. 2001. Podręcznik dla użytkowników Kwestionariusza Ogólnego Stanu Zdrowia, część I. In: *Ocena zdrowia psychicznego na podstawie badań kwestionariuszami Davida Goldberga. Podręcznik dla użytkowników kwestionariuszy GHQ-12 i GHQ-28*, eds. Z. Makowska, and D. Merecz, 13–189. Łódź: Oficyna Wydawnicza IMP.
Hasselhorn, H. M. 2003. *Working Conditions and Intent to Leave the Profession among Nursing Staff in Europe NEXT, Nurses Early Exit Study; a Research Project Initiated by SALTSA and Funded by the European Commission*. Stockholm: National Institute for Working Life.
Herzberg, F., B. Mausner, and B. Bloch Snyderman. 1959. *The Motivation to Work*. New York: Wiley.
Heszen-Niejodek, I. 1997. Style radzenia sobie ze stresem – Fakty i kontrowersje. *Czasopismo Psychol* 3(1):7–22.
Heszen-Niejodek, I., and H. Sęk. 2007. Zdrowie i stres. In: *Psychologia. Podręcznik akademicki*, vol. 2, eds. J. Strelau, and D. Doliński, 681–734. Gdańsk: Gdańskie Wydawnictwo Psychologiczne.
Hobfoll, S. E. 1989. Conservation of resources. A new attempt at conceptualizing stress. *Am Psychol* 44(3):513–524.
Hoel, H., C. Cooper, and C. Rayner. 1999. Workplace bullying. In: *International Review of Industrial and Organizational Psychology*, vol. 14, eds. C. Cooper, and I. Robertson, 195–230. Chichester: Wiley.
Hoel, H., S. Einarsen, and C. L. Cooper. 2011. Organizational effects of bullying. In: *Bullying and Emotional Abuse in the Workplace: International Perspectives in Research and Practice*, 2nd ed., eds. S. Einarsen, H. Hoel, and D. Zapf, 145–161. London: Taylor & Francis.
Hoel, H., and D. Salin. 2003. Organisational antecedents of workplace bullying. In: *Bullying and Emotional Abuse in the Workplace: International Perspectives in Research and Practice*, eds. S. Einarsen, H. Hoel, D. Zapf, and C. Cooper, 203–218. London: Taylor & Francis.
Hogh, A., and A. Dofradottir. 2001. Coping with bullying in the workplace. *Eur J Work Organ Psychol* 10(4):485–495. DOI: 10.1080/13594320143000825.
Hogh, H., E. Mikkelsen, and A. Hansen. 2011. Individual consequences of workplace bullying/mobbing. In: *Bullying and Emotional Abuse in the Workplace: International Perspectives in Research and Practice*, 2nd ed., eds. S. Einarsen, H. Hoel, and D. Zapf, 107–128. London: Taylor & Francis.
Kivimaki, M., M. Virtanen, M. Vartia, M. Elovainio, J. Vahtera, and L. Keltikangas-Jarvinen. 2003. Workplace bullying and the risk of cardiovascular disease and depression. *Occup Environ Med* 60(10):779–783.

Krejtz, I., and K. Krejtz. 2007. Wprowadzenie do analizy regresji jedno- I wielozmiennowej. In: *Statystyczny drogowskaz*, eds. S. Bedyńska, and A. Brzezicka, 364–384. Warszawa: Wydawnictwo Szkoły Wyższej Psychologii Społecznej "Academica".

Leary, M. R., C. E. Adams, and E. B. Tate. 2006. Hypo-egoic self-regulation: Exercising self-control by diminishing the influence of the self. *J Pers* 74(6):1803–1831.

Lee, R. T., and C. M. Brotheridge. 2006. When prey turns predatory: Workplace bullying as a predictor of counteraggression/bullying, coping, and well-being. *Eur J Work Organ Psychol* 15(3):352–377. DOI: 10.1080/13594320600636531.

Leymann, H. 1990. Mobbing and psychological terror at workplaces. *Violence Vict* 5(2):119–126.

Leymann, H. 1992. *From Bullying to Expulsion from Working Life*. Stockholm: Publica.

Leymann, H. 1993. *Mobbing – Psychoterror am Arbeitsplatz und wie man sich dagegen wehren kann*. Reinbeck: Rowohlt.

Leymann, H. 1996. The content and development of mobbing at work. *Eur J Work Organ Psychol* 5(2):165–184.

Matthiesen, S. B., and S. Einarsen. 2004. Psychiatric distress and symptoms of PTSD among victims of bullying at work. *Brit J Guid Couns* 32(2):335–356. DOI: 10.1080/03069880410001723558.

Mikes, P. S., and C. L. Hulin. 1968. Use of importance as a weighting component of job satisfaction. *J Appl Psychol* 52(5):394–398.

Mikkelsen, E. G., and S. Einarsen. 2002. Relationships between exposure to bullying at work and psychological and psychosomatic health complaints: The role of state negative affectivity and generalized self-efficacy. *Scand J Psychol* 43(5):397–405. DOI: 10.1111/1467-9450.00307.

Niedl, K. 1995. *Mobbing/bullying am Arbeitsplatz. Eine empirische Analyse zum Phänomen sowie zu personalwirtschaftlich relevanten Effekten von systematischen Feindseligkeiten* [Mobbing/bullying at work. An empirical analysis of the phenomenon and of the effects of systematic harassment on human resource management]. Munich: Hampp.

Nielsen, M. B., and S. Einarsen. 2012. Outcomes of exposure to workplace bullying: A meta-analytic review. *Work Stress* 26(4):309–332. DOI: 10.1080/02678373.2012.734709.

Nielsen, M. B., S. B. Matthiesen, and S. Einarsen. 2008. Sense of coherence as a protective mechanism Among targets of workplace bullying. *J Occup Health Psychol* 13(2):128–136.

Notelaers, G., S. Einarsen, H. De Witte, and J. K. Vermunt. 2006. Measuring exposure to bullying at work: The validity and advantages of the latent class cluster approach. *Work Stress* 20(4):289–302. DOI: 10.1080/02678370601071594.

O'Moore, M. 2000. Critical issues for teacher training to counter bullying and victimisation in Ireland. *Aggr Behav* 26(1):99–111.

Ogińska-Bulik, N., and Z. Juczyński. 2008. *Osobowość a stres i zdrowie*. Warszawa: Difin.

Pickett, C. L., and W. L. Gardner. 2005. The social monitoring system: Enhanced sensitivity to social cues as an adaptive response to social exclusion. In: *The Social Outcast: Ostracism, Social Exclusion, Rejection, and Bullying. Sydney Symposium of Social Psychology Series*, eds. K. D. Williams, J. P. Forgas, and W. Von Hippel, 213–226. New York: Psychology Press.

Quine, L. 2001. Workplace bullying in nurses. *J Health Psychol* 6(1):73–84. DOI: 10.1177%2F135910530100600106.

Raknes, I., S. Einarsen, S. Pallesen, B. Bjorvatn, B. E. Moen, and N. Mageroy. 2016. Exposure to bullying behaviors at work and subsequent symptoms of anxiety: The moderating role of individual coping style. *Ind Health* 54(5):421–432. DOI: 10.2486/indhealth.2015-0196.

Rodriguez-Muñoz, A., E. Baillien, H. De Witte, B. Moreno-Jimenez, and J. C. Pastor. 2009. Cross-lagged relationships between workplace bullying, job satisfaction and engagement: Two longitudinal studies. *Work Stress* 23(3):225–243.

Rodríguez-Muñoz, A., G. Notelaers, and B. Moreno-Jiménez. 2011. Workplace bullying and sleep quality: The mediating role of worry and need for recovery. *Behav Psychol* 19(2):453–468.

Salin, D. 1999. *Explaining workplace bullying: A review of enabling, motivating, and triggering factors in the work environment.* Working paper no 406, Helsinki: Swedish School of Economics and Business Administration.

Skogstad, A., S. Einarsen, T. Torsheim, M. S. Aasland, and H. Hetland. 2007. The destructiveness of laissez-faire leadership behavior. *J Occup Health Psychol* 12(1):80–92.

Strelau, J., A. Jaworowska, K. Wrześniewski, and P. Szczepaniak. 2005. *Kwestionariusz Radzenia sobie w Sytuacjach Stresowych. Podręcznik.* Warszawa: Pracownia Testów Psychologicznych PTP.

Tehrani, N. 2004. Bullying: A source of chronic posttraumatic stress? *Brit J Guid Couns* 32(3):357–366. DOI: 10.1080/03069880410001727567.

Tepper, B. J., M. K. Duffy, J. M. Hoobler, and M. D. Ensley. 2004. Moderators of the relationship between coworkers' organizational citizenship behavior and fellow employees' attitudes. *J Appl Psychol* 89(3):455–465.

Thomas, K. W. 1992. Conflict and negotiation processes in organisations. In: *Handbook of Industrial and Organisational Psychology, Part 3*, eds. M. D. Dunnette, and L. M. Hough, 651–718. Palo Alto: Consulting Psychologists Press.

Vartia, M. 2001. Consequences of workplace bullying with respect to the well-being of its targets and the observers of bullying. *Scand J Work Environ Health* 27(1):63–69. DOI: 10.5271/sjweh.588.

Warszewska-Makuch, M. 2007. Polska adaptacja kwestionariusza NAQ do pomiaru mobbingu. *Bezpieczeństwo Pracy – Nauka i Praktyka* 12:16–19.

Williams, K. D. 2007. Ostracism. *Annu Rev Psychol* 58:425–452.

Zalewska, A. 1999. Job satisfaction and importance of work aspects related to predominant values and reactivity. *Int J Occup Saf Ergon* 5(4):485–511.

Zalewska, A. 2003. "Skala Satysfakcji z Pracy" – Pomiar poznawczego aspektu ogólnego zadowolenia z pracy. *Acta univ lodz Folia psychol* 7:49–61.

Zalewska, A. 2006. Zadowolenie z pracy w zależności od reaktywności i wartości stymulacyjnej pracy. *Prz Psychol* 49:289–304.

Zapf, D. 1999. Organizational, work group related and personal causes of mobbing/bullying at work. *Int J Manpow* 20(1–2):70–84.

Zapf, D., and C. Gross. 2001. Conflict escalation and coping with workplace bullying: A replication and extension. *Eur J Work Organ Psychol* 10(4):497–522.

Zapf, D., C. Knorz, and M. Kulla. 1996. On the relationships between mobbing factors, and job content, social work environment and health outcomes. *Eur J Work Organ Psychol* 5(2):215–237. DOI: 10.1080/13594329608414856.

Zellars, K. L., B. J. Tepper, and M. K. Duffy. 2002. Abusive supervision and subordinate's organizational citizenship behavior. *J Appl Psychol* 87(6):1068–1076.

Index